Chaos Media

Chaos Media

A Sonic Economy of Digital Space

Stephen Kennedy

Bloomsbury Academic
An imprint of Bloomsbury Publishing Inc

B L O O M S B U R Y
NEW YORK • LONDON • OXFORD • NEW DELHI • SYDNEY

Bloomsbury Academic
An imprint of Bloomsbury Publishing Inc

1385 Broadway
New York
NY 10018
USA

50 Bedford Square
London
WC1B 3DP
UK

www.bloomsbury.com

First published 2015
Paperback edition first published 2016

Library of Congress Cataloging-in-Publication Data
Kennedy, Stephen.
Chaos media : a sonic economy of digital space / by Stephen Kennedy.
pages cm
Includes bibliographical references and index.
ISBN 978-1-62356-706-4 (hardback : alk. paper) 1. Mass media—Philosophy.
2. Digital media—Philosophy. 3. Mass media—Technological innovations.
4. Chaotic behavior in systems. I. Title.
P90.K4575 2015
302.2301—dc23
2014027508

ISBN: HB: 978-1-6235-6706-4
PB: 978-1-5013-2442-0
ePDF: 978-1-6235-6220-5
ePub: 978-1-6235-6724-8

Typeset by RefineCatch Limited, Bungay, Suffolk, UK

For Gemma, Lori, and Erin
And in memory of my brother Michael

Contents

Preface

This book is about humans in digital space and how they exist in a chaotic environment where time and space have become fractured and discontinuous.

It is about reality, and the real, the virtual, and the imagined. As such, it is a metaphysical enquiry into where we are. Yet it does not privilege the human. It recognizes and takes seriously the contribution of Actor Network Theory and the importance assigned to non-human factors in complex arrangements. In fact, from a Foucauldian perspective, it engages with an enquiry that describes the movement into the digital age in a manner that is mindful of Bruno Latour's methodology. It also recognizes the work of the speculative realists and object-orientated philosophers in this regard. Yet it is also a book that is mindful of acknowledging creativity, hence it owes a lot to Henri Bergson and Gaston Bachelard and their attempts to resolve some of the lingering questions around art and aesthetics.

This is a book about space, digital space. As such, it is concerned with boundaries, thresholds and borders. It is about the inside and the outside of space and place. Yet insofar as it recognizes the constantly changing nature of space, time is never far from its thoughts. How we came to inhabit digital space is an important question, and where this journey will ultimately take us is maybe even more important, but that will have to wait for another time, as it were.

Finally, it is a book about sound, about an analogical deployment of the sonic as a means of accounting for the ways in which a 'one substance cosmology' gives rise to specific events. One such event is the emergence of digital space itself, and as a way of exploring how this came to be, and to show how the variety of theoretical concepts on offer can be mixed into a meaningful solution, the specific sounds of both Coventry and Detroit will be triggered.

Introduction

> *The sociological theory that the loss of the support of objectively established religion, the dissolution of the last remnants of pre-capitalism, together with technological and social differentiation or specialisation, have led to cultural chaos is disproved every day; for culture now impresses the same stamp on everything.*
>
> (Adorno and Horkheimer 1997: 120)

Do we now live in a period when this chaos can no longer be so easily challenged?

We should be clear about the idea of chaos from the outset. When Theodore Adorno and Max Horkheimer questioned unfounded assumptions about cultural chaos, they were challenging instrumental rationality and the strange idea that where chaos was most visible, order and conformity actually prevailed. Where are we today in this regard? In the 1990s warnings of chaos, anarchy and official impotence in the face of untethered technological change were common. In actuality, such claims turned out to be no more true than at the time of Adorno and Horkheimer's writing. But what they meant by chaos was something different than the contemporary meaning being employed here. Chaos is not used to describe the anarchic disarray of digital environments but rather to try and show how patterning occurs, not as a fixed universal phenomenon, but as a relative continuity punctuated by difference.

Political and economic forces continue to try and impose order on chaos by paradoxically highlighting their own impotence, as will be demonstrated in what follows. But their grip is always tenuous. The best way to understand this is not to tether chaos but to set if free. To reconstitute the chaos that is routinely presented to us in a cultural sense, by reformulating it based on philosophical understandings of difference and non-representation – to foreground mystery and enter the shadows that are cast in contemporary mediated space.

What is the connection between sound (or what is referred to here as a sonic economy) and the digital environment? This also needs to be settled from the outset so as to avoid any confusion in the reader before laying out a more complex argument. In digital space the unreliability and uncertainty of visual dimensions relating to extension and proximity are accentuated. Quantifiable markers and dimensional measures such as height, width, depth, distance, etc. begin to lose their significance. With regard to place, such measures will be seen to be equally problematic and limited in terms of the amount of information they are able to communicate. It will be argued that qualitative dimensions and non-visual indicators such as sound and texture may be more useful tools in facilitating a critical engagement with the chaotic digital universe we currently inhabit. Even when visual representations prevail, as they do in digital space, it will be argued that they do so in an acoustic, mobile, untethered fashion that is more like sound than vision.

What is meant by the term digital space? Digital space is not regarded here as an alternative realm or in terms of a real/virtual dichotomy but as a lived experience of space that is facilitated and/or augmented by technology. It is a space characterized by the ability to traverse great distances at speed (in terms of physical movement and transport as well as disembodied communication), to retrieve information almost instantaneously, to share cultural reference points, and engage in activities of exchange (an economy) of either tangible or non-tangible objects. It is a space where our experiences are increasingly mediated – often through lenses or screens to access either the very big or the very small – and represented in terms that are both fixed and transferable/sensible. It is a space that cannot be viewed in its entirety and that can only be fully understood in relation to the constant movement and vibrations that are at its core.

The 'digital' prefix in the title is regarded as being as temporal as it is spatial in so far as it names a time period in which the argument for qualitative relativity can finally be made. What follows, however, is predominantly about space, and though time will feature at certain moments, full consideration of the progression from the analogue to the digital, and of time itself as a concept, will have to wait for a further volume. For now, suffice it to say 'digital space' is recognized as having a temporal dimension in terms of both speed and history. It is a mode of instantaneous transmission of data/information/content, and it also denotes an historical period.

The following, then, is an attempt to pull together some of the key ideas that can help us to orientate ourselves in digital space – a space that is constantly configuring and reconfiguring itself. But if it is doing it itself, what does that mean for the humans who inhabit this space? To answer this, we will journey through ideas that divide the world as it is and the world as it appears, into material and ideal realms respectively. We will encounter attempts to make sense of these ideas and the implications such attempts have had for our understanding of the world and our knowledge of it (epistemology). The questions, 'what do we know?', and 'how do we know?', will figure at all times. The journey will take us from Plato's cave to the inner reaches of a quantum universe, and outwards to the cosmos.

In Plato's 'Simile of the Cave' (Plato 1987), the light of the sun is proposed as a threshold beyond which the knowing subject cannot progress. Levels of reality are unfolded in a manner that suggests only a theoretical conception of Being. All that can be known is quantitatively present and visible. That which cannot be seen can never be known in any meaningful sense. It contained a visual bias that was reinforced by Aristotle: 'Above all we value sight because sight is the principal source of knowledge' (Ihde 2007: 7).

Plato divided the world into two: the world as we experience it; and an abstract realm that could never be accessed. As such, his thought represented a polarized dualism. Plato described a world where wisdom was characterized by a turning away from the 'real' world with its animistic view, and towards a world of pure abstraction that could only be comprehended as an idealized realm. It was in this idealized real that truth and knowledge resided.

This dualism was reversed/inverted with the rise of realism as a new period in the history of human thought. It was a period where the here and now took precedence over the abstract beyond. This was evident in relation to Marx's inversion of Hegel that will be discussed in more detail later. It has, according to Herve Fischer, 'been inverted once again' (Fischer 2006: 22).

> With the quiet revolution of computer technology at the end of the twentieth century, we witnessed the extremely rapid deployment of a new transcendental world: the digital replica of the real world. Since time immemorial, we have sought to escape from the real world and have been on a quest for the ideal, the absolute; we have aspired to the perfection, beauty, and intelligence of another far more vivid world. We have devoted ourselves to a poetic reverie, like the great mathematicians and apostles of digital languages.
>
> (Fischer 2006: 22)

For Fischer, the digital realm has become home to activities and expressions of the soul, the latest manifestation of the desire to escape the real. It is a place to which the mind takes flight from the real world and which includes art, science, the economy, politics, social issues, education and communication, television, film, and music, all of which are activities that are essentially human. However, the migration into the digital, he feels, throws up issues and dangers relating to posthumanism and the potential diminution of human influence. Such dangers represent a challenge, he says, for artists, researchers and philosophers and scientists. The challenge is to address the excesses of claims relating to the digital. This challenge is taken up here.

The reconstituted dualism of body and soul, material and ideal, raises complex issues in relation to the real and the unreal, the real and the virtual, the real and the imagined. The question is how to address such dualisms and the challenges they pose? For Fischer, among others, the challenge seems to be to resist the polarization and in so doing return the human and a degree of realism to the debate as a means of countering excess. This is one option. Another is to challenge the dualism altogether. This latter option is the one chosen here.

In order to do so, a number of philosophical positions will explored. After Plato, Descartes would re-order the split in relation to the mind/body duality. Leibniz and later Heidegger would try and do away with the split altogether (and largely this book seeks to maintain this collapsing of distinctions in relation to the real and the virtual). For Leibniz, monadological thinking proposed a dense material plenum consisting of ever-smaller parts that configured themselves – creating order out of chaos – and always in the best possible way. For Heidegger, the essence of Being was not to be abstracted from the real but evidenced through Dasein, that peculiarly human way of being that could be demonstrated via the development of a concentrated *way* of thinking capable of revealing essences. It was a way that got rid of Plato's dualism, and in doing so also collapsed the division between citizen and poet, or realist and artist. For Heidegger, the technical and the poetic were reunited under the term 'techne', a form of prehending the world that will be discussed in more detail below.

More recently, there has been a turn towards reinstating the dualism of the appearance and essence, best characterized by Quentin Meillassoux's argument in favour of a mathematical real that both precedes and precludes the human perception of it (Meillassoux 2012). In a world increasingly reduced to code and algorithmic calculation, it is not without merit. Along with Actor Network Theory and object-orientated philosophy, it does speak to some important questions raised by the posthuman debate.[1] But in the same way that Alfred North Whitehead and Henri Bergson needed to find a place for the individual and the concept of creativity in Leibniz's monadological universe, the argument being developed here similarly needs to find a place for those concepts in a contemporary digital landscape. In order to do so, the tendency to create divisions between the real and the virtual (remembering that for Deleuze via Bergson, the virtual is real, as will be shown later) or between citizen and artist will be resisted in favour of a singular world of quantum uncertainty. Such a world is not conducive to reliable visualization and instead we need to listen to fluctuating resonances and dissonances that move in, around, and sometimes through individual beings understood as temporary events in a chaotic universe.

To understand this universe adequately, then, will lead ultimately to an engagement with what Adrienne Janus calls the anti-ocular turn (Janus 2011). But why sound rather than vision? What is it about sound that can inform a methodology that is capable of accounting for a critical engagement with the contemporary technological age?

Sound, and listening itself, have been changed by advances in technology: 'our capacities for listening are changed by technological culture. Its roots lie in the birth of the electronic communications revolution' (Ihde 2007: 4).

We listen further, deeper, and in more detail with the aid of technology. This has meant a rejuvenated contemporary interest in listening, according to Don Ihde (2007). This is undoubtedly true. What is at stake here, however, is less how much technological advance has impacted on sound and listening, and more how a rejuvenated sonic discipline can itself serve to explain those technological phenomena – or if not explain them, provide a suitable critical framework for understanding them.

The Western philosophical tradition has consistently created a hierarchy of senses, with the visual at the top, that then proceeds to undermine the senses altogether. The result has been the establishment of a two-tier conception of reality where the ultimate reality is either reduced to an entirely abstract theoretical realm, as in Plato, or one where certain atomic constituents of reality (Democritus) or basic qualities of extension (Descartes) can be said to exist, because they are visible. Even then, such qualities as these are masked by other qualities that arise in the mind to form the richness of experience as subjective phenomenon. Even John Locke, as the advocate of empiricism, maintained the division between the objective and subjective realms. As did Nietzsche with his Apollonian/Dionysian split (Nietzsche, 1967).

Don Ihde provides a clear account of the primacy of the visual as a distinct feature of Western philosophy since Democritus and Plato. He sets out a trajectory that leads to a 'double reduction', by which he means that the senses themselves have been reduced to the visual, and that that prioritized sense is itself, along with all the other senses, ultimately devalued as a reliable indicator of objective reality. At the centre of the privileging of the eye is a more general distrust of the senses and the importance of experience – even in Locke. Where sense is recognized in the form of Democritean atoms or Cartesian extension, such perceptible data is ambiguous to the extent that it is cloaked by abstract thought and imagination so as to render the visible almost immediately invisible. For Descartes, vision can reveal extension – something is there, of that we can be sure, but what that thing is, is obscured by thought that triumphs over experience: 'The reduced abstract object (extended object) becomes "objective" and its appearance within perceptual experience with the significant exception of those ghostly remaining visual qualities becomes "subjective"' (Ihde 2007: 11).

So beneath our perceptual subjective experience hides an objective reality that can exceptionally be seen as a ghostly figure or outline that is coloured by the imagination. But how does this relate to a study on digital space? If we persist with the emphasis on the visual, it allows for a basic dualism between the real and the virtual to be upheld. However, 'It might even be preliminarily suspected that precisely some of the range of phenomena at present most difficult for a visualist tradition might yield more readily to an attention that is more concerned with listening' (Ihde 2007: 14)

This position was prefigured by Henri Lefebvre in his work, *Rhythmanalysis* (2004), in which he advocated a methodology capable of accounting for phenomena not conducive to visualization. He said:

> The electrical wire and the pole that runs alongside it say nothing about the energy they transmit. In order to understand, one must be an engineer, a specialist, and know the vocabulary, the concepts, the calculations . . .
>
> The same goes for our economo-political society. The visible moving parts hide the machinery.
>
> Is there nothing left of the visible, the sensible? Is our time only accessible after patient analyses, which break up the complexity and subsequently endeavour to stick back together the pieces? It is not necessary to go too far: a truth pushed beyond its limits becomes an error. The gaze and the intellect can still grasp directly some aspects of ours that are rich in meaning: notably the everyday and rhythms.
>
> (Lefebvre 2004: 15)

In order to expose the machinery, to go beyond the visible moving parts, an element of sonic, or acoustic engagement needs to be developed and followed. This is a task that is central to philosophical enquiry and the identification of primary and secondary qualities: that is, those qualities that are essential, objective, quantitative and present in the thing in itself, as opposed to those ideal subjective qualitative elements that are projected onto the thing of itself by sentient beings. But how far should we go in this quest? As the argument develops, it will be important to think about this question. Lefebvre's augmented methodology, and Nigel Thrift's hybrid approach (examined in Chapter 3) are in stark contrast to Quentin Meillassoux's outlining of a reality accessible only to mathematical formula, for example, as will be discussed in more detail below.

In order to address Lefebvre's question of whether there is anything left of the visible, the argument will oscillate between talking about sound, as waveform, as vibration and movement, and about music more specifically. And when we are discussing music specifically, it will be important to distinguish between music as a mathematical method of organization and as a less formal affective energy force – as discrete particle or continuous wave. The intention is not to choose between these options but to think about the usefulness of each. As Susan McClary has stated:

> The tendency to deal with music by means of acoustics, mathematics, or mechanistic models preserves its mystery (accessible only to a trained priesthood), lends it higher prestige in a culture that values quantifiable knowledge over mere expression, and conceals the ideological basis of its conventions and repertories. This tendency permits music to claim to be the result not of human endeavour but of rules existing independent of humankind. Depending on the conditions surrounding the production of such a theory, these rules may be ascribed to the physical-acoustic universe or may be cited as evidence for a metaphysical realm more real than the imperfect material social world we inhabit.
>
> (McClary, cited in Attali 2006: 150)

Again, the question, what is the role of the human? Also of importance here is the question of how one distinguishes between the 'real' and the 'more real'? For McClary, these are questions that relate to music, but they also relate to the broader task at hand here of exploring the status of humanity in the digital age. But it is not a case of simply moving these questions from one context to the other but of harmonizing them so that sound – and that includes music – can serve to augment our understanding of digital space. It will do so by addressing the debate around correlationism that has been central to the turn towards speculative realism and object-orientated philosophy that is discussed in Chapter 1.

Related to this argument and to the role of the human more generally, research recently designed to take the human out of the creative process in terms of musical composition offers up some interesting findings, and it is worth pausing for a moment to consider this in relation to what has been said so far.

A story on the BBC news website recently reported: 'You might think that creating the perfect piece of music – whether it's a classical great, jazz masterpiece or pop hit – is all down to the composer's talent, flair or even genius.' Not so, according to Armand Leroi, Professor of Evolutionary Developmental Biology at Imperial College London, who said: 'What we are trying to find out is whether you need a composer to make music.' Leroi's instincts were that you didn't. He believed a much more fundamental force of nature was at work.

'We don't often think of music as evolving, but everybody knows it has a history and it has traditions. But if you think about it, it really has evolved, it is changing continuously,' Professor Leroi explains:

> There are all the same forces of change, variation, selection and recombination as different musical traditions join together, transmute and fuse and divide again. This is all the stuff that is familiar from our understanding of the biological world, but we see it here in music as well.

He added: 'We believe music evolves by a fundamentally Darwinian process – so we wanted to test that idea.'

In order to do so, Leroi was joined by Dr Bob MacCallum, Mosquito researcher at Imperial College London and together they created DarwinTunes. The idea behind it was simple: to see if music could evolve out of noise – without the controlling hand of a composer (http://www.bbc.co.uk/news/science-environment-18449939) [Accessed 10 Sept. 2014].

These sonic processes that 'join together, transmute and fuse and divide again' can be seen as operating in a manner akin to biological systems, but do they also mirror political and economic systems? This will be examined in Chapter 2 in relation to the movement of statements and discourses that contributed to the formation of digital environments, and in Chapter 4 in relation to John Holland's Echo system that modelled economic phenomena (Waldrop 1992: 260). The biological analogy will also be invoked in Chapter 5 where the music of the Detroit techno collective Underground Resistance will be described as a 'mutant gene' (Goodman 2010: 156).

In an academic paper outlining the research, Leroi said:

> The music made by the world's cultures is immensely diverse. Because music is transmitted from one musician to another, and frequently modified in transmission, this diversity must arise from descent by modification rather like the diversity of living things, languages, and other cultural artifacts. What drives this process? It is often supposed that the music we listen to is primarily the product of aesthetic decisions made by 'producers' (i.e., composers, performers). Early Greek texts speak of specialist composers/performers, and the rudiments of formal musical theory, at least 2,500 years ago, and specialist composer/performers are found in many other societies as well. However, the reproduction, spread, and persistence of particular songs must also depend on the preferences of 'consumers' (i.e., the people who listen to them). These preferences are also clearly a selective process and, like any selective process, can have a creative role. Disentangling the roles of composers, producers, and consumers in shaping musical diversity is difficult in existing musical cultures.[2]

This 'arising from descent', coupled with the economic realities of distribution and consumption represents a complex model of analysis that can account not just for the production of music but the operation of digital environments more generally. It is a model that will be called upon as the argument in this book unfolds. It is a model that can also be related to the concepts of repetition and generality as set out by Gilles Deleuze in his *Difference and Repetition* (2008). Repetition and generality, Deleuze said, must be distinguished in order to account for the different economic conditions that subsist in each. Repetition is conducted from the heart, generality from the head. Repetition is a singularity that cannot be exchanged and has no equivalent – its currency is gift or theft. Generality consists of resemblance and formal equivalence – its currency is value and exchange. Repetition is Dionysian acoustic, generality Apollonian and visual. These aspects combine in a sonic economy, as they do in Leroi's model to constitute complex arrangements that defy observation in the classical sense.

Arising by descent implies gift or theft, as repeating singularities collide. Once quantifiable value is assigned, however, the notion of exchanges enters the equation, and more formal economies begin to prevail. This is the case in metaphorical terms in so far as it seeks to explain wider social phenomena, but it is also apposite in relation to the world of computer models that strive to account for such complexity.

As Deleuze has said:

> If repetition exists, it expresses at once a singularity opposed to the general, a universality opposed to the particular, a distinctive opposed to the ordinary, an instantaneity opposed to variation, and an eternity opposed to permanence. In every respect, repetition is a transgression. It puts law into question, it denounces its nominal or general character in favour of a more profound and more artistic reality.
>
> (Deleuze 2008: 3)

The point of developing an argument along these lines is to ask what form a profound artistic reality might take. It is also to affirm an approach that can account for invisible phenomena and that includes energies, repetitions and other phenomena that are often marginalized. In Chapter 5, this will be demonstrated as the motor cities of Coventry and Detroit are examined from this perspective.

Leroi and MacCallum's DarwinTunes worked using an algorithm to maintain and model a population of digital genomes, each of which would encode a computer program:

> When a program is executed, a short, seamlessly looping polyphonic sound sequence, a loop, is produced deterministically. Each genome/program specifies note placement, instrumentation, and performance parameters; however, tempo, meter, and tuning system are fixed for all loops. No human-derived sounds, rhythms, or melodies are provided as input to the algorithm. During the experiments, loops periodically replicate to produce new loops. The daughter loops are not, however, identical to their parents for two reasons. First, in a process analogous to recombination, the genome of each daughter loop is formed from the random combination of its two parents' genomes. Second, in a process analogous to mutation, each daughter also contains new, random genetic material. These two processes mimic the fusion of existing, and invention of novel musical motifs, rhythms, and harmonies that can be heard in musical evolution. The only selective pressure in DarwinTunes comes from a population of consumers who listen to samples of the loops via a Web interface and rate them for their appeal. These ratings are then the basis of a fitness function that determines which loops in a given generation will be allowed to mate and reproduce. We therefore

> expect that the frequency of musical traits will evolve under the influence of this selective process rather as trait frequencies in organisms do under the influence of natural selection.[3]

The rating process, or the allocation of a value to repeatable singularities gives rise to a mixed economy of repetition and generality. Such an economy, it is claimed here, can serve as the basis for an understanding of contemporary digital space.

In an earlier article, Leroi and Swire suggested that organic and cultural diversity have in common the fact that they are both the product of what Darwin called 'modification by descent' (Darwin 1882). They argue that both genes and cultural attributes are transmitted from one person to another, and that the transmission is often imperfect, so much so that they change with time. They go on to say:

> Of course, where organic traits are generally transmitted from parent to offspring by the rather restrictive system of Mendelian inheritance, cultural attributes have far more complex, less restrictive, and poorly understood, rules of transmission (Cavalli-Sforza and Feldman 1981; Boyd and Richerson 1985). Nevertheless, a logical consequence of any modification by descent process is that the distribution of diverse objects—species, genes, languages, pottery shards, songs—observed at any single time will contain traces of that history. The question, then, is how do we recover it?[4]

The idea of song as an object that is transmitted as part of complex cultural exchange is what is here named as a sonic economy. This will be explained in methodological terms more clearly below and will be used in the final chapter when musical phenomena will be employed in relation to the wider social economic and political transformations that took place in Coventry and Detroit. It also relates very closely to the notion of the event as discussed by A.N. Whitehead, and this will be further elaborated in Chapter 3.

Leroi and Swire were testing the usefulness of comparative data about songs compared with location-specific factors – as will be done here in Chapter 5 – to assess how they could facilitate findings in relation to the movement of past populations. Questions arose as to whether common factors were necessarily contiguous, or whether they pointed to a common origin dispersed through the movement of human populations. In doing so, they drew on the Cantometrics of Alan Lomax (Lomax *et al.* 1968). Lomax had developed Cantometrics to create a formal methodology capable of statistically analysing and quantifying song difference. That difference existed between specific locations and cultures was clear, but how did those differences interact, transmit, exchange, fuse, etc.? Leroi and Swire worked with the original Lomax data set that categorized 5,189 songs across 801 cultures [classified according to Murdock's *Ethnographic Atlas* (1967)] and which drew up modal characteristics that could be statistically organized to show song-style patterns.

They commented:

> When we looked at the Cantometric dataset we quickly rejected the idea of replicating Lomax's study. Innovative for their time, his methods had been rendered obsolete by forty years of advances in statistics and computing power. We also rejected reconstructing the history of individual musical lineages using conventional phylogenetic algorithms rather as has been done for languages (Gray and Jordan 2000; Gray and Atkinson 2003; Atkinson and Gray 2005; Dunn *et al.* 2005). Such algorithms assume that the evolving entity—be it a song, a repertoire, a language, a gene, or a species—evolves both slowly and independently of its congeners. One or both of these assumptions are clearly not met in songs which, due either to multiple invention or borrowing (in the terms of population genetics, recurrent mutation and recombination), show a bewildering mix of attributes in which geographic and historical signals are only dimly discernible.

In not conforming to conventional assumptions, musical lineages were capable of producing findings that studies in other areas were not. Geographic and historical signals were only dimly discernible – hence the need for a sonic economy.

The key significance of Leroi and Swire's article is that it raises very interesting and relevant questions about statistical, quantifiable evidence as opposed to qualitative effect. In a discussion of Victor Grauer's work, they are critical of his lack of statistical evidence and his reliance on listening as a research methodology.[5] They find no definitive evidence in terms of common descent as it relates to song style. There is no evidence to prove, they say, that song-styles that are deemed to be similar did not evolve independently of each other. There is also a lack of evidence in terms of why certain song styles occurred in certain locations. This requires that we take account of the difference between homologous (common descent) arguments and analogous (independent) ones.

In doing so, Leroi and Swire turn to the ideas of Jonathan Stock: 'Stock (2006) provides three lines of argument against the global homology of P/B style. He argues that P/B style may have evolved many times because it is: (1) not as complex as it seems; (2) understood by pygmies and Kung in different ways; (3) resembles the duetting of any mother and her infant.'[6] For Stock, the same musical style can occur in different places, as the result of accidental factors, and not necessarily because they are carried by common ancestor. In terms of song style, this is significant because socio-economic resemblances between Coventry and Detroit produced very different sounds. This means resemblance is problematic and is why what follows concentrates on difference that repeats.

Whether one agrees with Grauer's homology or Stock's analogy, two things are important here: the use of music to understand broader issues in human history and the mathematization of characteristics using digital data modelling to make sense of chaotic environments.

Listening, or the liberation of the auditory, rather than replacing the visual is intended as a means of developing an immersive sensual engagement with the world that can reveal truth hidden to the visual. In many ways it constitutes a reinvigorated approach to both scientific and philosophical enquiry; one that dispenses with the theoretical and imaginative tropes of metaphysics. In their place it calls for a stylistics of engagement based not on static representations and generalities, but mobile experience where data is processed on an ongoing basis and is a reliable measure of an ever-changing reality that is as temporal as it is spatial.

This is why Phenomenology is important within the context of the work: the real/virtual dualism is challenged in favour of a unified Being where no distinction is made between objective and subjective experience. As such, it relies heavily on Heidegger's affirmation of Being and his resolution of metaphysics as truth is revealed without recourse to visual verification.

When we look at digital space, to what extent do we distrust what our eyes tell us? To reverse the 'double reduction' of the senses, we should not only augment the visual with the other senses but we should also affirm that in terms of a sensible analysis of digital space there is nothing to see – but it is there, we do experience it.

Again it should be stressed that what is meant by digital space is not just the virtual space beyond the screen but a unitary existential space that shifts constantly and which defies conventional visual logic, a space where entities take up superpositions – of location and dislocation.[7] This position is presented in contrast to Quentin Meillassoux's reaffirmation of the Cartesian dualism that separates the world as it is from the world as it appears.

Western visual bias served to shape and then interpret what it shaped to the detriment of other concerns. This has now got to the point where it has atrophied and cannot account for contemporary concerns. The following will show how it has failed to account for the nuances of industrial decline and how a sonic economy can begin to make amends.

Chaos Media: A Sonic Economy of Digital Space re-thinks our contemporary understanding of the digital age from five perspectives: scientific, political, aesthetic, economic, and cultural (the technological resides in all of these categories as will be demonstrated), and ties them together to form a coherent whole in order to demonstrate how critical thinking can be reconfigured using a methodological approach that uses 'chaos', 'complexity' and sound as systematic tools for studying contemporary mediated space and the technology that has come to characterize it. The significant reference points cited in each case could be taken as objects existing independently and in their own right, as Graham Harman might have us do. This option is not taken here. Rather they will be played together to make audible the complex relational ontology that exists as noise becomes music and chaos becomes ordered.

In *The Question Concerning Technology* (1977), Heidegger urges the reader to go beyond any material manifestation of technology if its essence is to be revealed. He challenges prevailing instrumental definitions by way of a discussion of cause and effect wherein the traditional fourfold model that dates back to Aristotle can be re-thought or, more accurately, remembered. The established cornerstones of causality, *causa materialis, formalis, finalis* and *efficiens*, later interpreted as matter, aspect/eidos/idea, telos, and logos, must, Heidegger argues, be treated not in isolation – a tendency that has seen the *efficiens* or logos prevail to the almost complete detriment of *finalis*/telos in modern times – but in unison, each informing the other *ad infinitum* and where, if there is to be any dominant factor, it should be the much-neglected *finalis*/telos. This means, if Heidegger's lead is followed, that particular attention should be paid to the ends that are set as part of an agenda in the name of technology. Such ends in turn, have two aspects – poiesis and *physis* – which in their own way reveal truth, the former via the intervention of an external body and the latter as part of a natural and internally driven unfolding. Poiesis brings into unconcealment that which can be and as a term is itself embedded in another – *techne*, that names both the creative and technical modes of unconcealing that in the original sense of the word combined in a relationship of mutuality to describe a stylistics of engagement that was concerned not just with the fact that a certain phenomenon came to be, but how and in what way it came to be. In contemporary usage, the creative and technical modes have become separated. This separation of meaning, with the technical definition having come to the fore, has led to the neglect of such stylistic concerns. What attempts there have been to reintroduce a stylistic element into the technological realm as a characteristic of the *new media* environment, have been largely superficial and a considerable gap between the creative and the technical remains.[8] The word *techne*, Heidegger pointed out, was also closely aligned to that of *episteme* in terms of how one knows and what one knows. This has also been neglected in contemporary usage along with the idea that knowledge itself is subject to processes of production that are themselves subject to historical and existential modification. A Heideggerian approach then facilitates a broadening of the concept of technology beyond the material to include an assessment of the discursive practices that accompany its development and which furthermore are central to its essence. The relationships that exist between words and things are interpreted from this perspective as playing themselves out across a plane of existence that is both inter-causal and mobile.

It follows then that one must speak of a discourse on technology that is both responsive and prescriptive in a manner that is multiple and dispersed.

This notion of dispersion released Michel Foucault and those who followed from the chains of fundamental determinism and essentialism: that discourse cannot easily be located and must instead be identified in different forms in different dispersed locations is the point that demonstrates the very significance of discourse as a category. Like the fractured and placeless networks whose constitution it is implicated in as part of this analysis, any useful understanding of the concept of discourse needs careful analysis in order to try to understand how statements group together to create an attractive force that is both aesthetically compelling and materially instantiated.

What is at stake here is our ability to account for the innumerable ways in which circumstances that are often taken as a given, actually came into being. This requires that conditions of actuality be assessed with recourse to complex sets of ideal and material interfaces. From words to things and back, in a flow that requires perpetual critique,'the finding of visibilities' involves identifying discursive formations that give bounds to things (Deleuze 1995: 96). This involves not only the identification of words and things and the causal relationship between them, but also the inter-causal correlative relationship that moves constantly back and forth.[9]

From this perspective, technology, as it is being discussed here, is not a specific object or collection of objects (Kittler 1995), but is a constellation of generative and regenerative elements in a number of forms that are not always stylistically uniform or thematically stable.[10] The concept of technology is never considered as fixed or constant but as fluid in historical, scientific, artistic, and philosophical terms. This means that any meaningful engagement with technology must embrace social, political and aesthetic aspects of its coming into being. Science and technology, rather than being abstracted from 'life', must be positioned as central to it as discourse permeates the spaces between formal and informal output. With the appropriate theoretical framework in place, it is possible to demonstrate how the correlative relationship between the digital network environment and related discourse as a set or group of statements linked across and within time and space, forms a materiality that does not need to make the distinction between ideal and material, hardware and software (Kittler 1995). Instead the focus shifts to interconnected multimedia forms of expression that in some way contribute to the make-up of the contemporary media environment where correlation replaces cause and effect and where at times things become unstable. Hence new discourse formations come forward as predecessors ebb out of time with the conditions of their existence.

The contemporary or 'new media' landscape might be described in simple terms as a space where real and virtual worlds collide. Such a statement obviously needs further investigation. This book will proceed from the digital turn of the 1990s before stepping back to explore how our orientation in space and time is perceived to have altered. To name something as *new*, however, and attempt to define its characteristics in contrast with its predecessors too often leads to it being rendered static, knowable and subject to categorical boundaries that seek to serve as exemplary criteria. *New media*, in contradistinction to old media, might be described as interactive, digital, or more generally as being a form of communication or a means of access to information that uses technological advances in computing. In doing so, however, many important nuances are lost, overlooked or simply misunderstood and a form of technological determinism is allowed to dominate.[11] To rectify such failings, this book will investigate a particular period in the recent past – not as static but as mobile and fluid – where a series of technological phenomena came into view, aligned in a way that appeared revolutionary for a moment and cohered with a complex network of related statements before fading away from the *new* to become the *now*.[12] This transformation will be shown to have been indicative of a discontinuity or a tailing-off of a discourse in terms of its strength and significance before its regeneration and return to the kind of prominence currently being witnessed.

It will propose the idea that in order to fully understand any such alteration, an acoustic or sonic mode of thought must be developed to replace the dominant visual, representational, and instrumental logic of the past. Such positions fail to engage with the existence of mystery, preferring to present and re-present the world as transparent, visible and ordered. In a manner that reflects Robert Paul Wolff's demystification thesis in *Money Bags Must Be So Lucky* (1988), this work will show how the contemporary landscape is instead a chaotic, often invisible place, shrouded in mystery where echoes resound in the shadows.

Whereas visual cultures/economies are prefaced upon quantifiable and linear equivalences of cause and effect, sonic/aural/acoustic are descriptors that support a model for organizing chaotic elements in a sophisticated non-linear and non-representational way. They contain at their core the potential to recognize and account for the current state of fractured dissonance, that can at times flip into resonance and harmony, without imposing closure and ontological certainty. Equally they facilitate an expanded critical engagement with questions of territorial demarcation and political resistance that make up the specific case study in the book.

Overview of the book

Specifically the book will address the following points.

Chapter 1 [Science] Sonic Dimensions: This opening chapter will move from Descartes, via Locke and Leibniz to Nietzsche and beyond to include Newton and Einstein, in order that the idea of space as relative be extrapolated. This will be done in such a way as to critique the dominant visual representation of space. In its place, a sonic non-representational mode of thinking will be initiated: one that takes seriously the simultaneous spatio-temporal compression and expansion that characterize the contemporary digital age. This is important in terms of our ability to engage socially and politically at a time when both community and individual identity are being reconfigured within virtual communication environments. It will also show how sonic materialism works at both the philosophical and scientific levels of perception. Using Chaos Theory as a recurring motif (refrain), the ways in which digital channels of communication create complex connections and feedback loops to form new kinds of intelligence and social spaces will be explained.

Chapter 2 [Political] Territories of Resistance: New and emerging forms of political engagement will be examined using the Deleuzian concept of the 'refrain'. This concept organizes sounds together, bringing order to chaos. It consists of organized rhythms and patterns; daily routines and habits, the justification for which has long been forgotten. It marks out the territories that we recognize as our own or as belonging to others. It creates a sense of familiarity and belonging yet can also bring about alienation and a sense of detachment. It is our local neighbourhood, our political system and our state. Music, in certain forms, according to Deleuze and Guattari (1987), can de-territorialize this ordered terrain – disrupting the rhythm to create a momentary energy field that distorts the dominant order of things. The argument will be made that the 'refrain' is a useful theoretical tool in terms of organizing dispersed phenomena into a coherent form. This will be worked through via an examination of the discourse in relation to communication network technologies that began to emerge in the mid-1990s, and the proto-constitution of digital space.

Chapter 3 [Aesthetics] Echostate: This chapter combines Michel Foucault's category 'statement', with Gaston Bachelard's (1994) use of 'reverberation' and 'echo' in his introduction to *The Poetics of Space*, and expands their scope by examining how statements are echoed and amplified across a range of media that now exist in multiple simultaneous forms. Such myriad aesthetic forms are repeatable but not always identical in as much as they can take on different digitally enabled guises. It is the interaction between these varying forms that creates the dominant technology agenda of the digital age. The more general intention is to draw some conclusions about what came to be known as the *new media information age* – as a term or phrase and also as a specific aesthetic phenomenon situated within a wider discourse about technology that was itself multiple in nature and which served to establish a particular normative framework in scientific, political, aesthetic, economic, and cultural terms. It will also examine the concept of noise as a virtual plane that is the ground for emergent events.

Chapter 4 [Economic] An Invisible Exchange: Proceeding from a discussion of mystery and demystification in relation to economic systems, this chapter will outline the parameters of what is meant by the term *sonic economy*. As industrialism morphs into an information society wherein the dispersion of discrete elements is present but not always clearly visible, an appropriate dynamic, mobile mode of analysis is required. Such an analysis must engage with phenomena that are not easily seen. Robert Paul Wolff's thesis that modern economics and capitalism are largely regarded as devoid of mystery and, as such, viewed as plain and visible will be set out in a clear and relevant fashion. He argues, on the contrary, that they are rich in mystery and need to be carefully investigated if they are to be fully understood. The work being outlined here echoes this belief and proposes a methodology capable of achieving it. In doing so, it will also engage with the work of the Santa Fe group and their use of complexity and Chaos Theory in understanding economic phenomena.

Chapter 5 [Cultural] Motor Cities: This chapter – situated between the industrial and information ages – will use the examples of Coventry and Detroit to show how sound as an organizing principle, and music specifically, can help to explain social and political phenomena during a period of decline in traditional manufacturing and the rise of a technological aesthetic. It will argue that connections between the two cities based on resemblance are largely meaningless, and that if any connection is to be made, then it must take place at the level of difference.

Notes

1 The idea of posthumanism has been well debated and will not be repeated here. It has been covered in detail by: Donna Haraway, 'Situated Knowledges' in *Simians, Cyborgs, and Women* (New York: Routledge, 1991); Katherine Hayles, *How We Became Posthuman: Virtual Bodies in Cybernetics, Literature, and Informatics* (Chicago: University of Chicago Press, 1999); and Randolph Rutsky, *High Techne: Art and Technology from the Machine Aesthetic to the Posthuman* (Minneapolis, MN: University of Minnesota Press, 1999).

2 See http://www.pnas.org/content/109/30/12081.full [Accessed 10 Sept. 2014].

3 See http://www.pnas.org/content/109/30/12081.full [Accessed 10 Sept. 2014].

4 2006. See http://www.armandmarieleroi.com/wp-content/uploads/2012/08/Leroi-and-Swire.pdf [Accessed 10 Sept. 2014].

5 Victor Grauer, 'Echoes of Our Forgotten Ancestors', *the world of music*, 48(2) (2006): 5–59.

6 http://www.armandmarieleroi.com/wp-content/uploads/2012/08/Leroi-and-Swire.pdf [Accessed 10 Sept. 2014]. P/B style refers to Pygmy/Bushman style, generally thought to be the oldest form of music (http://music000001.blogspot.co.uk/2007/06/20-pb-survivals.html [Accessed 10 Sept. 2014]). The criticism of homology is from: J.P.J. Stock, *World Sound Matters: An Anthology of Music from Around the World* (London: Schott, 1996).

7 Referring to the extent to which the concept of time resists mathematical treatment, Henri Bergson has written: 'Its essence being to flow, not one of its parts is still there when another part comes along. Superposition of one part on another with measurement in view is therefore impossible, unimaginable, inconceivable' (Bergson 1992: 12).

8 R.L. Rutsky's *High Techne* (1999) and a large amount of popular commentary in relation to digital media technology share a technological aesthetic where creativity has become a quantifiable commodity that atrophies in direct proportion to its own reproducibility. Such attempts to reconnect poeisis and *techne* are superficial in so far as they concentrate on the *look* or on received notions of creativity wherein *efficiens*/logos continues to prevail.

9 The description of the archive as audio-visual is important in justifying the wide range of sources/statements that are drawn upon in order to identify a discourse. What is being proposed here is not a false consciousness where technology is misrepresented but an always contingent materiality in a permanent state of formation.

10 Heidegger prioritizes the discursive ground in *The Question Concerning Technology* to the extent that the concrete materiality of the environment in which linguistic practices take place are marginalized or at least reassigned to their 'rightful' place in a hierarchy of meaning. Foucault does something similar in *The Archaeology of Knowledge* (2004a). Kittler's view of materialism with its emphasis on hardware at the expense of software is a useful counterpoint, as is his expansion of Foucault's archived statements, but he ultimately over-emphasizes the visual and does not account for the invisible or the sonic as important factors in discourse formation. Kittler asserts that Foucault has forgotten to consider the nature of the medium as technology in his discourse analysis, and that while he is correct in his analysis of literate culture with all its attendant implications and effects, it is not adequate for a post-literate culture. It is, however, not a question of hardware or software in the battle for material significance but the way in which the interaction between the two combines to create materiality. If Foucault does not adequately account for the nature of the medium and consequently did not recognize that this oversight rendered his method sterile in a post-literate world, as Kittler suggests, he does allow, given his refusal to give examples of statements for fear of hypostatizing them, for the use of his method, albeit updated, in a manner that can and should take account of contemporary media forms. Deleuze and Guattari reinvigorate the material as words and things interrelate within spatialized realms, milieux or territories where what is evident to them is a

complex multi-sensual system where patterns emerge as assembled organized forces (political, technological, economic, cultural, aesthetic) that are always open to reassembly and reorganization. They are less engaged with specifics of hardware and software and more with the formation of mood and a kind of cultural materiality.

11 There have been recent attempts at categorical definitions. John Carey and Martin C.J. Elton state: '[A] media service or product is new from the point when the underlying concept starts to be implemented until the service or product can be regarded as established in the marketplace (in the sense that the product or service is very unlikely to be withdrawn as a failure)' (Carey and Elton 2010: 1).

12 The word *moment* is used here in direct reference to Henri Lefebvre:

> For Lefebvre, moments are significant times when existing orthodoxies are open to challenge, when things have the potential to be overturned or radically altered, moments of crisis in the original sense of the term. Rather than the Bergsonian notion of *durée*, duration, Lefebvre was privileging the importance of the instant. The moment has a long tradition in Western thought, most recently in the writings of Kierkegaard and Nietzsche. For Lefebvre, it is above all Nietzsche's writings that are important. In Nietzsche's *Thus Spoke Zarathustra*, the moment, the *Augenblick*, the blink of an eye, is a gateway where past and future collide, and the image of the eternal recurrence.
>
> (Elden, cited in Lefebvre 2004: x)

1

Sonic Dimensions

All media organize and manipulate time and space in one way or another. And it is this mediated manipulation that is the main concern of this book. Lived experience, and the things that matter as part of that experience, are subject to change as sensory existence is augmented (not replaced) by technology. This was recognized most explicitly by Harold Innes (1951), and was later developed by Marshall McLuhan (1975). From the primacy of visual space to the anti-ocular turn – or from representation to non-representation – the importance of a trajectory that has repositioned the subject in relation to the object world needs to be assessed in relation to the contemporary digital environment and this will require the clarification of certain terms and concepts to allow us to move forward in subsequent chapters.

This chapter will examine the implications of a reinvigorated discussion of space, and will revolve specifically around the difference between visual and acoustic space in terms of dimensionality. It will also point towards a later discussion (in Chapter 3) of the work of the artist Nicolas Schoffer in terms of digital creativity and spatio-temporal possibilities that might be better expressed as opportunities to address the previous limiting effects of time and space.

Media is understood here as a complex amalgam of processes and technologies of producing, communicating and storing information. It also concerns the fundamental relationship between subject and object, and as such a number of philosophical positions will need to be set out and explored in order to orient ourselves in what is being referred to here as digital space. The central aim of this book is to make clear to the reader arguments that deal with the individual subject and their experience of the 'real world'. Obviously the concept 'real world' itself needs some investigating, and this will happen in this chapter. For a brief summary before proceeding: beginning with the ancient Greeks, questions have arisen with regard to what happens when individual subjects encounter external objects or phenomena. Such questions have tended to revolve around the concept of essence. Can a subject ever really experience the object of perception as it really is, in its essence, or are all sensory experiences subject to mediation to the degree that what is experienced is the thing *of itself* rather than the thing *in itself*? Or the thing as it *appears* rather than as it really *is*? Questions such as this have occurred and re-occurred, from Plato, to Kant and Descartes, and Locke and Leibniz; from Heidegger to Foucault and beyond. What has always been at stake in their investigations, has been the status of knowledge and the reliability of perception. Hence it has always been about representation also, and the form that appearance takes. So as mediation and representation become technologized once more and our existential space is reconfigured, the arguments need to be revisited. The key task will be to establish how such arguments might assist us in critically engaging with this reconfigured space. Digital space is here taken to be both space and place. Not a dichotomy of real and virtual but a unified lived experience where those two realms collide to create mediated experiences and environments without essential qualities. This is not the world of *The Matrix*; not a world spatially divided into the desert of the real and a simulacra.[1] The philosophical task is not to penetrate some outer core to expose the real, but to describe what is present, and to do so in a way that acknowledges, and at times challenges, the futility of that task in terms of its ability to reveal a transcendent truth. The drive for truth is also a drive to bring order, to bring chaotic uncertainties under control. In doing so, a number of possibilities present themselves: to accept our inability to access the transcendent realm and with it the limitations of our sensorily mediated experience; to impose reason and the power of imagination (metaphysics) to constitute reality in its ideal form, to finally dispense with metaphysics altogether and challenge the abstract conception of Being and instead prove its existence in essence, or accept the uncertain nature of material conditions and content ourselves with detailed descriptions of the places and situations we find ourselves in. If the latter path is chosen, does that mean submitting to chaos? What follows is an attempt to find out.

All of these possibilities and the respective avenues of enquiry that they open up connect with recent philosophical developments in the areas of speculative realism and object-orientated philosophy. The key goal of these approaches is to remove the emphasis on the human and place it on an objective reality that is itself uncertain and open to speculation. As we encounter these arguments below, it will be useful to note the difference between virtual and augmented reality, as spaces of lived experience. Whereas the former alludes to a spatial bifurcation between real and virtual, or the real and the abstract as discussed above in relation to Plato, the latter is concerned with a single space relative in construction where human and computer interact. So once more the question of whether our existential world should be divided into separate incompossible realms, or universalized as a singularity needs to be considered.

To do so will require us to engage with Heidegger and his attempts to bring the task of philosophy to a close and finally reveal Being. To do so required a reconfigured approach: it was an approach that asked of those who aspired to it, to demonstrate a commitment to the idea of contrivance. This meant orchestrating and arranging phenomena in unison. An example is the coherence of the four aspects of causality that Heidegger insisted must be implemented simultaneously – played together in harmony. It was an insistence that was present also in Deleuze's work where the commitment was honoured in the form of melodic lines, and in the process philosophy of Alfred North Whitehead. It is also present and honoured in the work of Jean-Luc Nancy who talks of a 'fading into permanence', as sonic lines of enquiry supplant the visual. This book continues this line of argument and proposes that the characteristics of the digital age make it more urgent than ever that such a line of enquiry is pursued. All previous attempts it seems have fallen on deaf ears. The intention here is to amplify and remix those previous calls in a manner that makes them both accessible and timely.

Universal and relative space

Space and place: the debates around what constitutes each are numerous and complex. What is at stake in such debates is the distinction between an immanent material order and an idealized form created to suit certain ends. This was the debate between Marx and Hegel, and later between Castells and Lefebvre. It is a debate that revolves around questions of materialism and idealism, around concepts of discernible reality and infinite virtuality. To fully understand the debate requires that we examine the differing positions relating to space as propounded by Gottfried Wilhelm von Leibniz and John Locke. This will operate as the starting point for a wider discussion of space and an assessment of positions that relate to either universalism or relativism. It will also include the idea of time, as an historical concept as well as a scientific one, and the ideas of Zeitgeist/time period, moment, linearity and nonlinearity.

How do we think about space, and more importantly why does it matter? The thing to stress at the outset is that, depending on the kind of perspective adopted, different conclusions can be drawn about a range of questions that are relevant to the task at hand here: how do we understand the environment or milieu in which we are operating, and what factors should be taken as significant in such an analysis? In short, what matters?[2] In answering this, the turn from absolutism to relativism is central. Absolute systems beget predictability and confidence in the fact that we can know how things will turn out. Relativity, on the other hand, throws much of that certainty into question. This was recognized and debated by Einstein and the quantum scientists that accompanied and succeeded him, and it has also been recognized by social theorists engaged with questions of historical patterns and trajectories. 'Foucault recognises the simple but telling "fact" that [the] phenomena, events, processes, and structures of history (however we may define them) are always fragmented by geography, by the complicating reality of things always turning out more or less differently in different places …' (Philo, cited in Thrift and Crang 2000: 209). This turning out differently is of great significance here and will be further developed in Chapter 5 where events classified as similar in terms of certain taxonomies produce quite different effects, and raise questions relating to difference and repetition, predictability and uncertainty, universal or particular histories.

Isaac Newton and Pierre Gassendi saw space as absolute. For them, universal laws could be established and reliable predictions made.[3] It was a view that regarded the universe as a gigantic mechanical system that, once all of its constituent parts were identified, could be understood in its entirety. Despite its brilliance and undoubted utility, problems with it slowly came to the fore. René Descartes was one of the first to attempt to address some of its perceived failings. He tried to combine absolutism and relativism by distinguishing between internal and external place: internal place as absolute, and external place as relative (Casey, 1998: 162–3).

Not until Leibniz, however, was the case for space as relative appropriately made. But before Leibniz made his indefatigable case, John Locke was making claims in the name of relativism. His argument began with a critique of Descartes and the equivalence of space and corporeality. Descartes regarded solid bodies as occupying a fixed three-dimensional volume of space – the space inside the extended dimensions of the body being absolute. The extension of bodies in space was the only attribute of a thing/object that could be reliably discerned by the senses – so the eye, as discussed above, was trustworthy *to a degree*. Any essential or meaningful engagement with an object, however, was beyond sensory experience and organized according to reason. In digital space where extension itself is rendered almost nonsensical, even this relative privilege afforded to sight loses what reliability it had been credited with by Descartes. If this is the case, and we have retreated, as Hervé Fischer believes, into yet another abstract realm, are we left only with reason as a metaphysical way of understanding digital space? Not if we revisit acoustic sensory experience as an addendum to the visual we are not. But before we do that, let's return to the question of absolutism.

Locke began his critique of absolutism by questioning the association between solidity and extension and their usefulness in thinking about space and place. He alerted us to the idea that things can shrink and expand – altering extension without altering the extent of penetrability, hence any relations between inside and outside are also strictly relative. Space, therefore, is not confined within certain limits, it does not need boundaries of impenetrability for us to conceive of it: it can consist of nothing and as such exists as 'pure space'.

For Locke, *place* was not the volume of space occupied by a body, but a coordinate *in* space that is always in relation to other coordinates determined by two-dimensional distance in relation to at least two other points. This means that we can understand places in terms of distance relations to other places – either static or in motion, fixed or mobile. For our purposes, what is interesting is how those coordinates operate in relation to time and speed and functioning communication exchanges as other dimensions, that in contemporary digital space must come into play (specifically, Coventry and Detroit in relation to each other but also in relation to Kingston and Düsseldorf respectively, but more of this later). So places, for Locke, are always points in space rather than being significant in terms of the space they occupy. They are significant in terms of their relationship to other points. Hence, any understanding of a place must always reach out – any purely internal review must always be limited in scope and usefulness. Hence Locke facilitates our travelling outwards to form connections. But at this stage such connections are still very much allied to visual coordinates.

The concept of place, for Locke, is not natural but is determined by human activity in relation to certain practical ends. It is designed to meet specific utilitarian needs. Such needs will be of central importance later in this book when the case studies are laid out. For now, however, it can be seen as a mathematical function that concerns itself with very specific measurements as reliable representations. As such, Locke follows the ontic logic (as termed by Heidegger) of Plato and others in between at the expense of other types of thought. 'Moreover, just as Descartes and Galileo both removed such merely "secondary qualities" as colour and texture and temperature from place, so Locke also discounts such qualities, given that none of them can be converted into calculable distances'(Casey 1998: 165).

To these 'secondary qualities' sound can be added to create a nonrepresentational account of place more conducive to contemporary conditions.

So Locke has aided us in distinguishing between space and place and in determining that the significance of a place must always be seen in relation to other places. Yet the status of space itself, or 'pure space' remains ambiguous and open to charges of absolutism and voidness – an empty in-between. Sound cannot travel between a void – so Locke, though useful up to a point, does not take us far enough. In order to develop the argument, the spatial thinking of Leibniz needs to be added to the mix. Leibniz shared with Descartes the idea of space as a dense material plenum. This is important in relation to the case being made here as it enables the accommodation of sound, and the removal of distance as a significant factor in distinguishing between distinct places (Leibniz, 1992, *Monadology* #57).[4] Distance applies only to things that are extended in space, and implies limits or two-dimensional extensions as Locke would have it. But what if a thing, a city, for example, was not reducible to its mathematical extension, its quantitative vectors? What if it was a compound of non-extensive non-representational components? If this were the case, it could be explained as a sonic event with temporal phenomena combining, dispersing and recombining constantly.

The downgrading of distance as a reliable measure of place is an idea that lends itself extremely well to a critical engagement with contemporary technological environments or media ecologies where connections across a material plenum (wired and wireless) form complex interacting systems (Fuller 2007). Such environments simultaneously replicate and communicate qualitative features of places in a manner that reduces distance to an almost meaningless factor.

Leibniz thought that distance in terms of both space and place had some uses but it did not tell the whole story. It was only useful, he thought, in pinpointing objects that were extended in space. However, objects themselves were, he thought, comprised of more fundamental elements (monads) that were not spatially extended and whose position in relation to others could not be measured in conventional metrics, given their metaphysical status. To be spatially extended and relatively located implied for Leibniz a visual verification of existence. But he was venturing beyond such common-sense judgements. So place for him was not equivalent to an absolute 'inside' with impermeable limits (*à la* Descartes), nor a set of simple coordinates (*à la* Locke). Instead it was a contrivance of qualities, limits and coordinates between them that might not be so conducive to measurement or visual identification. For Leibniz, space was not a container of specific sets of relativities between places, but was itself a relative concept in relation to innumerable potential situations, some of which come to fruition, but most of which do not. Either way, space conceived of in this way is always relative to the potential that inheres within it. So space is an ever-changing realm of possibilities marked by places or 'situations' that require qualitative as much as quantitative consideration if they are to be adequately understood. What might be becomes equally as important as what is – for the 'is' will always potentially be something else. 'Taken in its totality, the complete collocation of coexistent things is the order of space'(Casey 1998: 168).

How well does this describe digital space, or more simply space in an age where digital forms of production, storage and distribution proliferate? The collocation of existent things is altered as spatio-temporal relations are increasingly mediated by technology, and this is the case as much with regard to analogue technology as it is with digital. This spatio-temporal mediation will be explored in detail in Chapter 5. Coventry and Detroit are 3,669 miles apart but how *close* are they if we take close to mean something that can be measured by more than just distance?

In order to address proximity/closeness in a manner that is useful here, the idea of extension as a marker of place needs to be explored further. For unlike Descartes, Leibniz regarded extended beings as constitutive of their own place – and not of absolute space. Yet, for Leibniz, extension does not *define* the material body. A thing is always something more than its size and shape, but these other elements may not be themselves extended or verifiable in terms of perception – that is, not measurable or conducive to representation. What is extended in this regard is a quality rather than a quantity. So, extension is equivalent to place, but it is not a fixed quantitative place but a qualitative and relative one insofar as it is not subject to a measurable constant (how, for example, has Detroit relatively extended its qualitative reach beyond its quantifiable geographic position?). So places are diffusely located: qualities are present within a fixed (relatively) locus and from there emanate as continuous extensions of the simultaneously present elements that comprise the quantitative and perceptually compresent aspect of place. This movement out beyond compresence requires us to think of places as entities that will not and cannot be confined to a specific location that can be understood visually – in so doing, they become temporal as well as spatial, more acoustic than visual. This means that an entity, or an extended body is not a distinct self-sufficient configuration of distinctive parts with quantifiably definitive boundaries, but is rather a continuous phenomenon 'whose parts inhere in each other in a continuous series of overlapping members' (Casey 1998: 170).

There are strong similarities here to Henri Bergson's notion of time as duration, or a continuous sonic phasing of points on a chromatic scale whose beginnings and ends are indecipherable in flow. This raises a very important point in relation to continuous (analogue) views of time and space and distinct (digital) ones. The Bergson/Bachelard debate is central here in terms of assessing how digital space should be approached, and how it can inform the particular case studies in this book (there will be a more detailed discussion of time in Chapter 3).

Is digital space continuous or discontinuous? We must distinguish in asking this question between a historical sense of space as a temporal unfolding, and the practice of space – or practices in space that can be regarded as discrete, discontinuous, spontaneous and unpredictable – chaotic. What happens when the continuous diffusion of place is processed as a relatively compresent configuration of bits (or monads) that are themselves distinct and discontinuous? What happens to spatial relativity then? This is digital space: a continuous discontinuity of relations that are themselves not spatially extended. Rather, they are infinitely reconfigurable insofar as they constitute a space of what is possible as much as what is present and perceptually verifiable. This might also be termed quantum space or a space of uncertainty (Werner Heisenberg) where knowing where something is means you can never know where it is going and vice versa.[5]

Extrinsic and quantifiable differentiating qualities such as distance or location are not present in the thing itself, therefore, in terms of understanding that thing or place, they do not exist; they are imaginary. Instead the qualitative difference must be found *in* the thing itself – not between things. The importance of this observation as a combination of Leibniz and Deleuze rests on the extent to which extrinsic 'differences between' are made less significant or even pointless in digital space. This in-between is now a medium through which travel continuous overlapping qualities destined to meet with other qualitative vibrations that travel out in all directions to form connections.[6] They do so in a manner akin to Belousov Zhabotinsky's famous reaction, and the spontaneous formation of animal markings as accounted for by Alan Turing.[7] Again here we must be careful not to confuse digital space with its effects on real space. Instead the digital and the analogue 'inhere in each other in a continuous series of overlapping members' to repeat the quote above.

A qualitative relativity then seems the most appropriate way of engaging with questions of space at a time when dimensionality is impacted on by technology to such a significant extent as to finally require us to make the leap that Leibniz never felt fully confident enough to do. For Leibniz, in order for two entities, things or places to be different from one another, simple coordinates of position (quantitative matters) are alone not sufficient. There must be some qualitative difference also. Position is an extrinsic quality and insofar as it only has meaning in relation to other things is not present in the thing itself. Hence it is abstract and of only limited use. 'Paradoxically, "the interconnexion of things" is better served by quality than quantity, for quality alone possesses an "intrinsic accidental denomination"' (Casey 1998: 172). This seems like a better way in which to understand digital space wherein notions of extension and quantifiable relativity are of little if any importance. It is a space where 'intrinsic accidental denomination', or patterns of self-organization proliferate and become entangled as discontinuous and discrete events in a continuous material plenum of superpositional potentialities. So digital space is comprised of a series of interconnections that need to be understood as relative and qualitative phenomena.

For Leibniz, place is qualitative yet firmly located while space is the quantitative and relative colocation of places. But what happens when distance itself, the arbiter of relativity, becomes qualitative also? Or when a place is released from its locational point and allowed to float? Consistent advances in technology have allowed this to happen. What happens to place in digital space, or more appropriately here, what happens to specific place[s]? How are their qualities diffused and how far do they extend and how fast?

To try and answer these questions, the contribution made by Michel Foucault to our understanding of space should be considered.

Foucault and space

As phenomena, statements move in relation to one another across three realms of space: (1) collateral – which invokes the relationship between statements in terms of similarity and/or proximity/distance that gets over the problem of compresence in both spatial and temporal terms; (2) correlative – that assesses the relationship between statements and their subject; and (3) associate – that assesses the internal institutional logic of statements and the circumstances of their production (Deleuze, *The New Archivist*, in Deleuze 1988).[8] Only by employing each will the complexities of the current age be able to be accounted for.

The complexity of events and the uncertainty of outcomes were central to Foucault's work. He was aware of the spatial dimension of discourse/power and the inherent problems of attempting to visualize such relations, and this will be explored in more detail in Chapter 2 where the political aspects of technology and digital space are explored. '[W]hat is wholly visible is never seen in its entirety. It always shows something else asking to be seen; there's no end to it. Perhaps the essential has never been shown, or, rather, there's no knowing whether it has been seen or if it's still to come in this never ending proliferation' (Foucault 2004c: 112). For this reason, the visible is not an entirely reliable register. This is a point similar to the one put forward by Kevin Lynch in his book *The Image of the City* (1960). Drawing on themes that resonate with the task at hand here, he wrote:

> Looking at cities can give a special pleasure, however commonplace the sight may be. Like a piece of architecture, the city is a construction in space, but one of a vast scale, a thing perceived only in the course of long spans of time. City design is therefore a temporal art, but it can rarely use the controlled and limited sequences of other temporal arts like music. On different occasions and for different people, the sequences are reversed, interrupted, abandoned, cut across. It is seen in all lights and all weathers.
>
> At every instant, there is more than the eye can see, more than the ear can hear, a setting or a view waiting to be explored. Nothing is experienced by itself, but always in relation to its surroundings, the sequence of events leading up to it, the memory of past experiences.
>
> (Lynch 1960: 1)

This seems to describe perfectly the continuous discontinuity of space as an infinite set of possibilities that cohere as different discrete occasions for different people at different times.

Place then must be engaged with as a qualitative aspect of relative space not conducive to fixed visual representation or recognition. This is the key argument that is being developed in this book.

Foucault's critique of total history and the monolithic temporal blocs that it imposes is well documented. He was also interested, though, as Chris Philo points out, in critiquing the totalizing of space at the expense of 'smaller scale real differences and distributions' (Philo, cited in Thrift and Crang 2000: 211). So spatial as well as temporal continuity is at issue. Spatial continuity relies on a quantitative set of unchanging dimensions instead of the qualitative and 'accidental denominations' referred to above. To adequately grasp the relevance and importance of this requires that we reject the notions of resemblance outlined by Deleuze, and focus on specific, singular, local events.

But in doing so a problem is encountered: a crossroads, a decision to be made in terms of direction. The decision has often been described in terms of the debate between modernism and postmodernism. Philo refers to Derrida's version of the dilemma: the modernist dreaming in search of an origin or centre of intellectual enquiry as opposed to the postmodernist preference for 'play or chaos' (Derrida 1978: 292–3). This relates to Foucault's preference for the emergence of theoretical frameworks out of specific observations and not vice versa, for inductive rather than deductive thinking. Where, then, it must be asked, should the researcher focus their attention? On the universal or the particular? On the absolute or the relative? And here is another problem: the act of 'focusing' with its visual bias and its emphasis on bringing under control and fixing, an act that counters Foucault's 'never ending proliferation' in favour of borders and limits. So maybe 'focus' is the wrong word. Instead of focusing on certain specific aspects of any given situation or phenomena, perhaps what is required is a sensory immersion or a sonic journey through space, or what Henri Lefebvre has called a *Rhythmanalysis* (Lefebvre 2004).

In doing so, many phenomena can come into and fade out of play – not play in the frivolous sense but rather in the operational sense where everything is regarded as significant. This implies chaos in the scientific sense whereby even the most apparently insignificant phenomena assume great importance and command our attention. This is what happens in Foucault's genealogical method where discrete elements are carefully accumulated to build up ever more complex narratives, and is most explicitly expressed in his analysis of the work of the poet Raymond Roussel in *Death and the Labyrinth* (Foucault 2004c). Foucault describes this Rousselian space in terms of the coexistence of things large and small, with an absence of hierarchized phenomena. It is a space where juxtaposition reigns and where the revealing of essences is absent.

Despite exhaustive description and journeys through the myriad options available within this Rousselian labyrinth, true meaning beyond the simple existence of things never reveals itself – the thing itself remains silent for Foucault, and, by association, Roussel. In one passage, Foucault recounts Roussel's description of a label on a bottle of water that journeys deeper and deeper into the scene, revealing ever more detailed surface observation, but never quite arriving at a 'meaningful' end (Foucault 2004c: 99–123):

> Nothing of what is seen here is given nor can be given visually. This visibility is self-contained and offers itself to no one, drawing an interior celebration of being which illuminates it from head to toe for a spectacle without possible spectator. It's a visibility separate from being seen. Although access to it is through a glass lens or a vignette on a label or a letterhead, it's not to stress the interception of an apparatus between the eye and what it sees, nor to insist on the scenes lack of reality, but as the result of reduction, to place the act of seeing in parentheses and at another level.
>
> (Foucault 2004c: 107)

The question might be asked as to the extent to which Foucault's account of Roussel's description of the label on the water bottle constitutes a virtual reality? The framed two-dimensional label invites the viewer into another world in a manner similar to that offered to us when we are surfing the web through the 'glass lens' of a computer, following hyperlinks and juxtaposing multiple non-hierarchized elements together in a temporal flow – never experiencing all that is offering itself to us until we travel deeper and deeper into the fractal maze. What needs to be stressed here, however, is the extent to which Foucault follows Roussel's lead not to determine the existence of separate worlds: the real and the literary, or the real and the virtual, but to assert a methodology of spatio-temporal description where real and virtual exist as aspects of a single realm with variable dimensional characteristics (continuous discontinuity). This coexistence or compresence becomes more evident in digital space as virtual reality moves into augmented reality. With this move, space becomes a multidimensional phenomenon, a chaotic realm, relative in nature, complex and unpredictable in practice, yet repeatedly patterned in actuality.

This tension between unpredictability and the emergence of patterns was one that occupied Foucault. In response to Roussel's *How I Wrote Certain of My Books*, in which he identified the rules that guided his work: 'Foucault is left speculating whether it may be that there really are similarly unnoticed and silent rules out there – and just waiting to be specified shaping not just Roussel's other texts, not just all literary compositions in general, but the whole patterning of human reality on this planet' (Philo, cited in Thrift and Crang 2000: 218).

For Foucault, the site or space of analysis is not a neat and coherent one, not an orderly bounded absolute locale wherein things, events and phenomena are conveniently proximate. Rather the space of analysis is a dispersed space – as set out in *The Archaeology of Knowledge* – that contains disparate objects of study that take multiple forms and which have to be ordered in a non-hierarchical juxtaposing in order that sense can be made of them and their relativity:

> And yet the envisaging of a space of dispersion is not tantamount to saying that all there is in the world is a chaos that the researcher can do no more than celebrate, because Foucault clearly supposes that there *is* some order in the dispersion waiting to be discovered, but that this order resides resolutely *in the things themselves* and not in any order theoretically imposed from without.
>
> (Philo, cited in Thrift and Crang 2000: 219)

Similarly, the task in this book is to lay things out – the large next to the small, not as a quest to reveal hidden truth but as an acknowledgement of the arrangement and rearrangement of things in infinite and relative patterns of accumulation. This is a description of the world as it is, not pandering to imaginary representations of truth or realities beyond human perception. As such, it raises the question of chaos theory and the emergence of complex self-organizing patterns. If for Foucault such patterns were speculative, for us they may be evidenced in listening.

For Foucault, there was a double problem: the inability of vision to comprehend the whole picture, and the futility of vision in terms of prising the truth from the silent object (or objects relative to each other) of its gaze. But what if they are not silent, what if from the immediate silence audible patterns emerge (Maitland 2009: 157)? This idea will be revisited later.

If Foucault identified problems with the visual – settling on the *isness* of the surface – what might be the implications if we augment his thinking with sound, or visual space with acoustic space? Not to discover essential truths but emerging patterns.

In *Laws of Media*, Marshall and Eric McLuhan argue for a new kind of thinking that is capable of accounting for the prevailing cultural norms present in what they called acoustic space (McLuhan and McLuhan 1988). This acoustic space constituted for them a return to a state of affairs that had for centuries been displaced by visual space.

Laws of Media begins with the collapsing of traditional dualisms – hardware/software, physics/metaphysics – in a manner that attempts to resolve the Platonic polarization discussed above. Marshall and Eric McLuhan propose a system of thought and analysis that makes no distinction between artifacts or media as tangible things, such as computers (hardware) or intangible things such as theories, or laws of science or philosophical systems, poetry or music (software). All, they say, are artifacts, and all are human. Hence, for them, everything, whether corporeal or not, is open to analysis according to a defined set of laws that can reveal truth.[9]

They go on to identify two ways of discovering truth or establishing a set of laws with reference to Francis Bacon's *Novum Organum*:

> There are and can be only two ways of searching into and discovering truth. The one flies from the senses and particulars to the most general axioms, and from these principles, the truth of which it takes for settled and immoveable, proceeds to judgment . . . And this way is now in fashion. The other derives axioms from the senses and from particulars, rising by a gradual and unbroken ascent, so that it arrives at the most general axioms last of all. This is the true way but as yet untried.
>
> (McLuhan and McLuhan 1988: 4)

This might be describing the debate in relation to absolute versus relative space, or to more recent discoveries in relation to chaos. Where the first option imposes universal models which phenomena must be made to fit, the latter acknowledges that even in systems where every possible variable is known, there is still the potential for the unknown to occur. The first imposes totalizing *a priori* models or axioms of the kind Foucault was so keen to dispel. As the former of these two options it had given rise to the dominant position of visual space as a privileged realm of scientific enquiry that replaced the multi-sensual realm of acoustic space, consigning it to pre-history. In short, visual space was a realm that privileged a 180-degree line of sight and acoustic space was a realm of 360-degree immersion where all of the senses were active simultaneously and in union.[10] The transition from acoustic to visual space took place, according to Marshall and Eric McLuhan, in ancient Greece, and was a process that took thousands of years. Slowly, they say, from Aristotle onwards, the visual sense came to be the 'common sense norm'. However, what had taken thousands of years to establish took only a few decades to reverse: 'By the twentieth century, visual space was obsolesced in all fields and acoustic space retrieved. The mechanical paradigm, enthroned in the seventeenth and eighteenth centuries, was eventually replaced by a field mosaic approach' (McLuhan and McLuhan 1988: 39). It is worth expanding on this point. In their text, Marshall and Eric McLuhan reference Langdon Winner in a footnote to explain the 'field mosaic approach'. In a manner that points forward to the case study to be carried out in Chapter 5, they quote Winner as having said:

> [T]he mid-twentieth century has brought the eclipse of the machine as a model for everything under the sun. Too many recent developments in science and technology – quantum physics, relativity, modern chemistry and biology, the alloys, plastics, the transistor – simply do not match the two primary images of the older mechanical tradition: Newton's clockwork universe and the cog and wheel machine of nineteenth-century industry. Artifice has become more subtle. Many devices properly called machine are no longer truly mechanical. Even Lewis Mumford, who emphasizes the idea of society as machine, has changed his emphasis to something called the 'Power Complex.' What needs expression is the idea of a set of large-scale, complex, interdependent, functioning networks which form the basis of modern life. For this, 'the machine' will no longer suffice.
>
> (Langdon Winner, *Autonomous Technology*: 193, cited in McLuhan and McLuhan 1988: 39)

This is describing a shift from continuous to discontinuous space wherein the predictable regularity of machines gives way to a fractured, chaotic, rhizomatic network society that is often irregular and unpredictable.[11]

> Continuity pertains only to visual space, and therefore the phrase 'space-time continuum' is misleading and a contradiction. There is no 'continuum' in acoustic space, whether pre- or post-Euclidian: it is formed as a discontinuous and resonant mosaic of dynamic figure/ground relationships. Although the language used in discussing the new post-Euclidian spaces reflects an underlying adherence to the visual – or absolute – space model, the actual perception of the scientists is that of a mythic double-plot of space and time without connections. They maintain that 'there is no such thing as a fixed interval of time independent of the systems to which it is referred.' Space and time serve as grounds for each other. Werner Heisenberg illustrates the confusion inherent in their language: '[in modern physics], one has now divided the world not into different groups of objects but into different groups of connections . . . What can be distinguished is the kind of connection which is primarily important in a certain phenomenon . . . The world thus appears as a complicated tissue of events in which connections of different kinds alternate or overlap or combine and thereby determine the texture of the whole.'
>
> (McLuhan and McLuhan 1988: 40)

It is these connections and the 'complicated tissue of events' that *Chaos Media* is attempting to make sense of. Or if not make sense of, at least support the development of a kind of thinking that will help in this task. In doing so, it adopts an immersive sense-based approach, drawing on a Heideggerian phenomenology, or a Foucauldian discourse model, or McLuhan's concept of acoustic space. As such, it may be accused of correlationism – that is, the idea that distinct subject–object dualisms are collapsed into an indivisible set of feedback loops connecting perception and reality. This kind of thinking has been challenged by the turn to speculative realism and object-orientated philosophy, and the implications of such a turn should be considered in relation to the argument being put forward here.

Speculative realism returns to the arguments of Locke and Leibniz in order to overcome correlationalism and to reassert a version of reality, or of materialism, that is distinct and independent from the perception of it. The real world exists, according to their argument, in a dimensional state that is not available to the senses – thus speculative, not conducive to proof based on experience. It is transcendent. Spatially, this is significant insofar as it re-establishes the dichotomy between the material and the ideal, of objective and subjective worlds, and it revisits the primary and secondary qualities set out by Locke describing an objective primary reality, onto which is projected secondary qualities. The independent existence of a material world beyond perception that can only be speculated upon, and where non-human actors operate in infinite combinations to create the conditions of their existence might seem to unproblematically describe our digital present. But is it necessary to return to the divide? If we do, what are the consequences, in general and in relation to this book?

The mind/body dualism that correlationism tried to overcome has been revisited by Quentin Meillassoux. He says:

> In order to reactivate the Cartesian thesis in contemporary terms, and in order to state it in the same terms in which we intend to uphold it, we shall therefore maintain the following: *all those aspects of the object that can be formulated in mathematical terms can be meaningfully conceived as properties of the object in itself.* All those aspects of the object that can give rise to a mathematical thought (to a formula or to digitalization) rather than to a perception or sensation can be meaningfully turned into properties of the thing not only as it is with me, but also as it is without me.
>
> (Meillassoux 2012: 3)

The return to a Cartesian dualism sets out a two-world scenario – meeting at the interface where the inaccessible or transcendent real presents itself to the subject and where the relationship or correlation occurs to constitute experience. The real, however, continues to have an existence independent of that constitutive moment. So what it is that we eventually correlate with can develop outside of human agency. Such a scenario supposes a division between a chaotic world to which we have at best limited access (its existence being conducive only to mathematic or digital verification) and a sensible world of lived existence that is to a degree organized and ordered. But what kind of space, or spaces, is this describing in relation to the discussion being developed here?[12] It is one where the digital realm (as mathematical real) constitutes a 'meaningful' space with or without human habitation. It is an independent realm whose nature can only be speculated upon – a real, independent of human cognition. This is a view largely shared here, but what is not shared is the idea that humans are positioned outside of this space. It must be asked, therefore, what are the implications of such a space for our relational subjective real – how do the two worlds collide if indeed they do?

The digital world is, unlike Descartes' objectively extended world, non-extended and not susceptible to spatio-temporal restraints; it is self-organizing and self-perpetuating. Descartes' objective world was a framework over which was superimposed subjective qualitative features. All that could be known of it was that it was there. Meillassoux's objective world must by necessity be of a different kind – a kind of inverse *Matrix*. Penetrating the real (or the anti-real for Meillassoux) represents a twist in what is commonly understood as the real/virtual dichotomy. The real becomes the virtual and the virtual the real. The inaccessible realm of formulae and code are both objective and abstract at the same time, and formations that occur therein do so in a manner free from human interference as code.[13]

The idea of an independent reality that exists without human affirmation is an interesting one, and is common in the dystopian scenarios of Hollywood films where technology eventually turns on its former masters. It requires that we rethink the idea of technological neutrality and our relationship with non-human actors (Latour 2005). But what if the real is not inaccessible, just invisible – not just quantitative (mathematizable) but qualitative, and, moreover, what if the two spaces proposed by Meillassoux are not separate?

As a means of correcting what Bruno Latour calls the 'much abused metaphor of the Copernican revolution', Meillassoux's proposal takes humans out of the bigger picture and places them in their own peripheral world where myth and representation continue to reign. In doing so he re-establishes representational thinking and the idea of a chaos that is by necessity kept at bay. He takes as objective, a real outside world over which humans have no control and where, in order to fend off the threat of the unknown that this outside poses, humans create a world in their own image. The problem with this dualism is not that it devalues humans but that it inadvertently re-establishes them as sovereign in their own, albeit meaningless realm. The extent to which this realm is prefaced upon the defence of order against chaos supports the continuation of the Apollonian/Dionysian divide with all of its attendant political implications. For Nietzsche, these were different states within a universal space and their interaction was an essential aspect of interpreting the world. It was a view shared by Karl Popper:

> Thus, like Descartes, I propose the adoption of a dualistic outlook, though I do not of course recommend talking of two kinds of interacting substances. But I think it is helpful and legitimate to distinguish two kinds of interacting states (or events), physio-chemical and mental ones. Moreover, I suggest that if we distinguish only these two kinds of states we still take too narrow a view of our world: at the very least we should also distinguish those artifacts which are products of organisms, and especially the products of our minds, and which can interact with our minds and thus with the state of our physical environment. Although these artifacts are often 'mere bits of matter', 'mere tools' perhaps, they are even on the animal level sometimes consummate works of art; and on the human level, the products of our minds are often very much more than 'bits of matter' – marked bits of paper, say; for these bits of paper may represent states of a discussion, states of the growth of knowledge, which may transcend (sometimes with serious consequences) the grasp of most or even all of the minds that helped to produce them. Thus we have to be not merely dualists, but pluralists; and we have to recognize that the great changes which we have brought about, often unconsciously, in our physical universe show that abstract rules and abstract ideas, some of which are perhaps only partially grasped by human minds, may move mountains.
> (http://www.the-rathouse.com/2011/Clouds-and-Clocks.html [Accessed 10 Sept. 2014])

If humans are regarded as configurations that arise out of chaos as events, and then become part of a universal and continuing re-patterning, the return to the Cartesian dualism is not necessary. Instead, digital space, and the place of humans within it, should be understood as a series of infinite possibilities that are temporal and multi-dimensional, and as a ground for journeys full of encounters, where data matters and where noise resolves to harmonize. It is an environment where patterns form and where predictions are by no means certain but can be formulated endlessly. The digital space that is being described here is a space of Leibnizian accidental denominations that does not require a return to the Cartesian dualism to understand it. It is a space where the digital is an aspect or a dimension within an eternally combining and recombining set of potentialities.[14] Put simply, either everything is real or everything in virtual, it doesn't really matter which. What matters is the plural singularity that Popper describes, where everything, including human perception, is reducible to basic elements (monads/atoms).

To conclude: visual space is a representational realm that is essentially quantifiable – a material reality that even when seen as relative in the Leibnizian sense can be mathematically posited. It is a space where visual reference points can be plotted and identified as pictorial map, as star map, or as a Google map according to longitude and latitude. It is a space of largely predictable zero sum outcomes that are absolute in nature. It is a space that has persisted over time, whatever McLuhan may have thought. It is a space through which Manuel Castells travels easily, but where Henri Lefebvre finds only limited answers. It is a space marked by the related temporal regularity of clocks. But is it a space that is still recognizable? Or are its clean edges beginning to blur – to phase and feedback? Have we finally made the transition (or the eternal return as recursive loop) to acoustic space, with all the sensual and chaotic simultaneity that that implies? Is digital space an acoustic space characterized by diffusion, dispersion, and uncertainty wherein the notions of place and space alike, as well as time, no longer conform to quantifiable relative formulations? For some, digital space is the very apotheosis of quantifiability – the Leibnizian monad made manifest – the internet of everything, the world wide web of significance, the rhizome, the matrix. For Leibniz, extension was not space but place – not an enclosed three-dimensional volumetric that can be quantified, but rather a fixed point, in relation to other fixed points, from which was emitted or diffused qualitative characteristics that cannot be easily measured or represented. It was the relationship between these points that constituted space. The argument being made here is that having highlighted the qualitative nature of place in order to challenge Descartes' position of volume as space, Leibniz then retreated to the relative safety of place as a fixed point. So 'place' for him is that fixed point. But what happens if we amplify the qualitative characteristics, not to counter Cartesian logic but to affirm the uncertainty surrounding the fixed nature of place, or to put it better, to extend qualitative thinking from place to space itself – so that now even space is not quantifiable? This is what digital space is: not a realm separate from the real but a qualitative aspect of the real that mediates spatio-temporal relations.

Notes

1 Žižek, *Welcome to the Desert of the Real* (2001) and Baudrillard, 'The Precession of the Simulacra', in *Simulations* (1983).
2 Marx, K. *Theses on Feuerbach*, available at: http://www.marxists.org/archive/marx/works/1845/theses/theses.htm [Accessed 8 June 2014].
3 Popper, K. *Clouds and Clocks*, available at: http://www.the-rathouse.com/2011/Clouds-and-Clocks.html [Accessed 8 June 2014].
4 '# 57. And as the same city regarded from *different sides* appears entirely different, and is, as it were multiplied respectively, so, because of the infinite number of simple substances, there are a similar infinite number of universes which are, nevertheless, only the aspects of a single one as seen from the special *point of view* of each monad.' Available at: http://www.marxists.org/reference/subject/philosophy/works/ge/leibniz.htm [Accessed 8 June 2014].
5 See McEvoy, J.P. and Zarate, O. *Introducing Quantum Theory* (Cambridge: Icon Books, 2004).
6 See http://www.bbc.co.uk/news/magazine-26464396 [Accessed 8 June 2014].
7 *The Secret Life of Chaos*, available at: http://www.bbc.co.uk/programmes/b00pv1c3 [Accessed 8 June 2014].
8 Heidegger prioritizes language in *The Question Concerning Technology* to the extent that the materiality of the environment in which linguistic practices take place is marginalized. Foucault may also be guilty of this. Deleuze and Guattari reinvigorate the material as words and things interrelate within spatialized realms, milieux or territories where rhythms reconcile the discursive and non-discursive.
9 This is an argument that is also made by Graham Harman in his *Quadruple Object* (2011).
10 For a full discussion of visual and acoustic space, see McLuhan and McLuhan, *Laws of Media* (1988: 13–66).
11 So as the machine ceases to serve as a metaphor for visual space and as it ceases to dominate as a continuous emblem of the modern economy something else emerges: a discontinuity, to refer back to Foucault, an epistemological break. This

can be contrasted with the man–machine aesthetic that came to define Detroit techno, and the High-Tech uncanny as described by Rutsky in his *High Techne: Art and Technology from the Machine Aesthetic to the Posthuman* (1999).

12 For an interesting view on the relationship between nature/natural objects and computational code, see: http://www.dataisnature.com [Accessed 10 Sept. 2014].

13 See Kroker and Kroker's *Codedrift*: available at: http://www.ctheory.net/articles.aspx?id=633 [Accessed 10 Sept. 2014].

14 These artworks explore the infinite possibilities of a chaotic realm where self-organizing patterns emerge and fade as sound and echo, see: http://vimeo.com/60209340 [Accessed 10 Sept. 2014]; or Dromos http://vimeo.com/80542263 (The provocation of a generative accident) [Accessed 10 Sept. 2014].

2

Territories of Resistance

The transformative power of *new media* technology is an issue that is being re-constituted at the top of the academic, social and political agendas. Funding streams that encourage engagement with topics such as the 'Digital Economy' as well as broader attempts to name a new discipline, with 'Digital Humanities' the current front runner, point to a renewed enthusiasm after a period of quiet reflection. The topic has also captured the popular imagination once more with the relaunch of *Wired* in the UK in 2009, the BBC *Superpower* series from 2010, and more recently with the role of social networks in the Arab Spring and in the riots in the UK becoming the subject of much debate and conjecture.[1] All of this in spite of the fact that *new media*, as the term is often understood, have been in our midst for almost two decades.[2] It would seem pertinent to ask then, whether at this moment in time, the term is still a relevant and meaningful one. Does it name an age, a process, or a specific technological phenomenon or does it form part of a broader political agenda? *New media* as a specific term must be defined in relation to all of these respective aspects. In order to do so, the following will consider these questions and address the notion of the *new* in relation to both media and technology as a means of questioning its utility. In doing so, it will consider the discourse around *new media* as a mobile refrain that seeks to bring order to chaos.

This chapter will look specifically at the historical period (Zeitgeist) 1994–95 that was characterized by the euphoric celebration of media-related technology. The period that encapsulated the 'digital turn' that as a phenomenon had been developing for some time, as will be shown. In doing so, the period under investigation will not be interpreted as having been new in the conventional sense.[3] Rather, it will show how it created its own non-linear timeframe with simultaneous reference to the past, present and future, operating as an agenda that was both prescriptive and responsive. 'The zeitgeist, as Godel noted, has its own time and agenda' (Yourgrau 2005: 143).[4]

The term 'agenda' is used here as both a plurality denoting a bringing to order, a collection or set that comes together within an organized framework – used as opposed to the singular 'agendum' – and also as a singularity in terms of its modern sense, used to denote a particular purpose or end and implying intentionality. However, it is not a simple matter of choosing between the two in terms of how the word is used. Instead the task is to demonstrate the interplay between them: the two meanings of the word are at all times entwined. If this is taken into consideration, then it is possible to retain a political focus without necessarily being required to demonstrate specific intentionality in situations where the evidence on offer shifts along an axis of causes and effect rather than forming a sequential and continuous timeline.

The agenda, in both senses of the word was, according to a conventional reading, misconceived as having no particular end. Politics during the period in question was predominantly viewed as helpless and subordinate to a tide of scientific, technological and economic progress that looked set to overwhelm all that stood before it. What a closer examination of official policy statements and legislative statutes reveals, however, is that they served to circumscribe particular aspects of the 'technological' landscape during the period 1994–95 and that the British and US governments made a significant contribution in terms of technical specifications, architectural planning, implementation and utility. But political activity did not and does not operate in a rarefied atmosphere; it occupies a space where myriad statements and practices are produced and disseminated – named here as digital space where the 'tissue of events' that McLuhan and McLuhan (1988) describe, connect. Hence political contributions as statements need to be related to the kinds of discursive practices/statements/events in 'popular' form – mediated through film, TV, newspapers, books and magazines – that began to proliferate during the same period, and which contributed to the Zeitgeist and the casting of technology as a determining force. In order to do this, a theoretical framework needs to be set out that, in the first instance, pushes to its limits our understanding of what is meant by both 'technology' and 'politics', and furthermore deploys a model based on an analysis of discourse and its relation to the material manifestation of technology. The political realm is not an abstract self-contained structure with a simple cause-and-effect relationship to technology, but is itself embedded within a nexus of interrelated realms or spheres of activity, both discursive and non-discursive that contribute to what technology itself essentially *is*. It is this complex interrelatedness and the connectivity between statements that are central to the argument being presented here.

A detailed uncovering of the political policy landscape specific to the development of digital space in the mid-1990s, when viewed in relation to other less overtly political statements alluding to technology, can provide an understanding of how complex discourse formations shaped this important *moment* (Elden, cited in Lefebvre 2004: x) – not as simple cause and effect but as complex inter-causality. Such an analysis will have methodological implications for understanding the contemporary iteration of technological discourse in so far as it is extrapolated not as a legacy in the linear sense but rather as a reconfigured version of an agenda that adopts new (in the Bachelardian sense) and different forms.[5]

What happened during the period in question cannot be explained adequately using traditional means, however. The seemingly unstoppable introduction of technology into all spheres of life has to date defied all attempts to encapsulate it: it was political but was cited as having escaped established political control, it was clearly technological but that assertion is only useful if we expand what is understood by this term as technology itself; it was economic yet seemed to represent a destabilizing factor for traditional economic practices, it was cultural in terms of human impact but equally post-human in terms of potential, it was always artistic and aesthetic but in a way that strained to the limit questions around form and representation. It was neither temporally nor spatially specific, yet it was subject to specific forces in terms of development and use. For these reasons, an enhanced approach is required. The *moment* itself and the environment that it ushered in can only be explained, it is argued, by developing a form of non-representational thinking (Thrift 2008) that moves beyond the Habermasian system/lifeworld distinction to explain how a dispersed multiplicity of interrelated phenomena combined and cohered in ever more complex ways to create the materiality of what came to be the contemporary, but no longer (if it ever was) new, political economy in its digital guise.[6] Such thinking must be capable of coping with the perceived chaos and uncertainty that come with the so-called 'information revolution' and associated digital networks of the *new media* environment. Commensurately it must not separate the fields of politics, economics, science and aesthetics or the arts, as has traditionally been done (Heidegger 1977: 156–82), but combine them in innovative ways that can make sense of the chaos and its fractal patterning.[7]

Identifying statements that cohere to form a specific discourse supports the argument that the official policy process had a role to play in shaping technological factors that would later become significant, and provides a contrast with the explicit 'non-interventionist' view that will be described below. It also supports the argument that as constituent elements of discourse, statements move across space and time, configuring and reconfiguring in constantly differing ways as part of a sonic economy. Hence the discourse on *new media* was not itself new in the conventional and continuous sense that the old had been replaced. Rather, it configured itself in a contemporary form, containing echoes of previous discourses, and was discontinuous in so far as it consisted of re-energized rather than replaced elements (Bachelard 1994: xvi). Discourses, from this perspective, are interpreted as being in a constant state of flux and thus do not conform to conventional cause-and-effect arguments that were common during the mid-1990s that cited technology as the cause of a new digital age.

Statements relating to the UK government's propensity to act in respect of technology were in existence prior to the 1990s, and a brief engagement with them will be useful in order to show how they recurred as constituent echoes in the formation of future discourses:

> In the early and middle 1960s there were a number of developments in British government which seemed to form a pattern and to point to a new style of managing the economy, even a new style of politics in this country ... There was obviously a predisposition to action and intervention on the part of the government, often involving a readiness to assume entrepreneurial as well as regulatory functions.
>
> (Williams 1971: 52)

The pattern being described here points to a controlled relationship between the political process and technology. It is a pattern that, rather than being replaced in a straightforward fashion, was reconfigured to take a different form.

Harold Wilson in his Scarborough speech in 1960 had talked of the relationship between socialism and science. At the 1963 Scarborough conference, as Leader of the Opposition, he made the famous 'White Heat' remark that referred to government-inspired technological development. By 1964, Labour was in government and had established the Ministry of Technology and the Industrial Reorganization Corporation experiment. The Ministry of Technology became the 'biggest state directed complex of scientific and industrial power in Europe' (as quoted in the *Observer*, 27 November 1966, cited in Williams 1971: 53). Evidence of further direct involvement continued in the following decades: 'Since the early 70's a small group of MP's of all parties, including Kenneth Baker, Christopher Chataway and the close friend and aide of Mrs. Thatcher, Airey Neave, had been pressing government to involve itself more fully in micro-electronics and information technology' (Hollins 1984: 51).

In 1978, the then Prime Minister James Callaghan had been alerted to the potential of new information technology by BBC's *Horizon* programme, *The Chips Are Down*. As a result of this programme, the Central Policy Review Staff (CPRS) was set to work on a number of reports, and to establish the central machinery required to properly control the 'new' technology issues (Hollins 1984: 51). The establishment of such a body was significant in so far as it demonstrated the relationship between official (government-level) statements and popular (mass media) examples that came together as part of a discourse. It is important here because it highlights the complex relationships at play in the area that might broadly be termed technology. While it is not being claimed that technology is always determined by political will at the point of invention, it can be demonstrated that the technology in question was subject to specific political processes in terms of appropriation and utilization (Mackie, cited in Heap *et al.* 1995: 41–53). In reacting to popular statements formulated for TV, James Callaghan set in motion a process that would see government more explicitly involved in future developments. It was a policy of intervention that was, however, largely unsuccessful. In terms of its stated intentions, it was also a policy that would change with the coming to power of the Conservative government in 1979.

If there was an ideological shift in the UK in 1979, and there clearly was, any assumption that there was a wholesale disengagement from technology, as part of a more general movement away from state intervention, is erroneous. All the pieces had not yet come together, but the technological environment that would later become visible as the digital space being explored here was beginning to be lit up by statements that would resonate and echo loudly in the years to come. As such, these pieces that had not been visualized might be described as virtual elements awaiting actualization. The specific relationship between the virtual and the actual will be discussed below.

In 1982, significantly designated 'The Year of Technology' in the UK, there was a cautious enthusiasm for the developments in IT and the emerging cable industry that was envisaged as being best placed to provide the requisite infrastructure. Such cautious enthusiasm was based on the principles of convergence that were being facilitated by advances in computing and the wider process of digitization. It was, however, an area of specialist interest. Debate was punctuated by excitement and tempered by critical engagement predominantly taking place at the academic level (Negrine 1985: 10). This enthusiasm, despite the illusory and apparent lack of governmental engagement, was consolidated throughout the decade that followed until a greater and more audible excitement began to surface in the 1990s as technological developments themselves, as well as opinions about them, began to resonate louder and in a more audible 'public' fashion. Such statements that could be heard were not necessarily saying the same thing about a particular subject/theme at the same time in the same place – not always easily connected collaterally – but they were in the process of cohering to form the ground that would eventually support the visible manifestation of the *new media* environment. There were statements with explicitly political themes, broader socio-cultural or existential themes, and economic themes. And despite there not always being a clear lineage or connection between them across time and space, each statement combined with others to form a discourse that would eventually consolidate a 'mind set' in relation to technology in a manner akin to Heidegger's concept of enframing in so far as that concept takes account of the whole contrivance of factors that finally contribute to what technology is (Heidegger 1977).

To fully understand how such contrivances work, consideration must be given to specific statements to show how the political process, despite professing to be entirely neutral in relation to technology, actually contributed to its eventual manifestation in terms of what came to be seen. Evidence of this can be found in the Command Paper, *Creating the Superhighways of the Future: Developing Broadband Communications* (CM 2734, HMSO 1994).

This document set out the government's position in relation to developing broadband networks and associated technological advances. An examination of the details of this particular report, chosen because of its content as well as its timing, is set against the background of the dissemination of a more popular discourse as they were seen to have been developing during the same period which claimed governments were increasingly unable to keep pace with technological advance.

The paper began by stressing the need for reliable broadband networks if the new information and entertainment services were to develop successfully; it did so in a manner that further reflected its ideological commitment to the market: 'The government considers efficient infrastructure is best developed by competing providers, rather than by promoting a single all-purpose switched two-way infrastructure' (Command Paper 1994: para 12, p. 4).

It also highlighted the extent to which its programme of liberalization had created an environment conducive to such development:

> [T]he government believes the UK enjoys a real advantage over its main international competitors: because of the regulatory and policy framework which was put in place when privatisation began in 1984, and was further developed in 1991 as a result of the review of the telecommunications duopoly, a process began of investment in new broadband networks, and of investment in the upgrading of existing networks.
>
> (Command Paper, 1994: para 4, p. 1)

When placed within the theoretical framework discussed above, an examination of such statements begins to show how an emerging discourse served to fundamentally affect the nature and shape of the technology in question, and to do so in a way in which those effects can be seen to have been strategically managed wherever possible at the political level.

The paper was organized into three sections: (1) the potential uses of the networks; (2) the role of government; and (3) the regulatory environment. Beginning with potential uses, the paper was keen to stress that opportunities went beyond simply multichannel TV. As an example of the increased awareness of this it cited: 'The fact that the cable operators have recently changed the name of their trade body from the "Cable Television Association" to the "Cable Communications Association" reflects the growing awareness of these opportunities' (Command Paper 1994: para 10, p. 4). By citing the 'awareness' demonstrated by an external body, the government depicted a kind of new reality that needed to be named or renamed in order to shift the emphasis from television to the broader category of communication. In doing so, it positioned itself outside of the environment in which change was taking place. It represented itself as a neutral observer when in fact it had been contributing to the creation of that environment for some time.

As if to suggest that the said opportunities had only then become apparent, the paper claimed: 'The full scope of opportunities offered by the convergence of telecommunications, broadcasting and information technology is only now becoming clear' (Command Paper 1994: para 8, p. 4). In reality, the 'full scope of opportunities' had been known since at least as early as 1982 as was seen in both the report published by the Information Technology Advisory Panel and the Hunt Report.[8] Both reports had demonstrated a clear vision of the potential for integrated and convergent communication systems. Yet it was a vision that was contradicted by the government policy of separation in terms of telephony and broadcasting, and within broadcasting itself between terrestrial satellite and cable provision that had served to restrict its development. This is an example of the misinterpreted relationship between official statements and seemingly unrelated external events. In practice, visibilities were found as the process of interaction between interrelated realms of activity cohered to bring the situation to light (Deleuze 1995). This was, however, not a simple cause-and-effect relationship. Politics, economics, science, media in all their forms, discursive and non-discursive factors alike formed complex interactions between dispersed statements that might be said to have been *in formation* or engaged in a process of coming together that was underway but not yet complete.

The paper further envisaged that: 'The services made possible by broadband networks for households are likely to be as significant as the first introduction of television ... choices will encompass not only entertainment but a wide variety of new services making passive consumers active demanders covering the whole range of household activities, including entertainment and education, work and shopping' (Command Paper 1994: para 20, p. 6).

This vision was illustrative of a discourse based on the private utilization of a technology aimed at consumers rather than citizens, one that concentrated on commercial rather than public concerns (Coleman, cited in Axford and Huggins 2001: 110). This relates specifically to the problems posed by Hervé Fischer in relation to what he calls Ultraliberealism (discussed in more detail in Chapter 4). In terms of possible public uses for the new technologies, the report limited itself to a description of the Super Joint Academic Network, and a brief outline of the possibility of extending the use of information networks within the NHS (Command Paper 1994: pp. 7–9).

It then moved on to set out its position on the role of government itself. Typically playing down its importance, it saw the role of government as peripheral; as being able to articulate a vision, set an example, and keep abreast of technological developments (Command Paper 1994: para 34, p. 10). It went on to say: 'In the future, the relative importance of regulatory issues is likely to recede, as competition becomes even more established in communication services' (Command Paper 1994: para 37, p. 10). However, its role had been far greater than that. Any reduction in the importance of government should be seen here as self-imposed rather than as a response to technological imperatives as others claimed (Angell 1995).

The paper went on to set out the government's position on what it termed 'A Stable Regulatory Framework' which made clear the government's decision to adhere to its own earlier recommendations contained in the White Paper, *Competition and Choice: Telecommunications Policy for the 1990's*. In Chapter 5, it looked at 'Further Government Responsibilities', beyond the regulatory framework. It stressed the government's commitment to 'working in dialogue' with those involved in the development of network technology (Command

Paper 1994: para 63, p. 16). Thus it unwittingly highlighted its own role in the discourse process. It referred to its efforts to 'support and develop', citing two examples where it had done this: Telecoms 2000 initiative – a series of seminars set up by the DTI, BT, Cable & Wireless among others to 'promote a message of continuous improvement', and a study co-sponsored by the DTI and the Cable Communications Association, looking at the future software requirements of the industry.

The paper also stressed – referring to a consultative exercise implemented by the CCTA (Central Computer & Telecommunications Agency, set up the previous November) – the need for government to explore new ways of using the developing technologies to undertake public business and to provide 'services to consumers' (Command Paper 1994: para 71, p. 18). The use of the term 'consumers' was indicative of the way in which the discourse process had, over the period under consideration, served to instil normative principles in the area of network technologies with free market liberal ideas.

Still further evidence of the extent of government involvement came with its pronouncements on the issue of research and development: 'An important role for the government is to help shape the content of UK and EU funded research and development programmes on communications' (Command Paper 1994: para 77, p. 19). 'The communications R&D programme will be known as Advanced Communications Technology and services' (Command Paper 1994: para 78, p. 19). 'In discharging this role of maximising the opportunities provided by research and development programmes, the government hopes that an important contribution will be provided by the Communications Technology Foresight Panel, organized by the Office of Science and Technology' (Command Paper 1994: para 82, p. 20). What this showed, despite the continuing reinforcement of its non-interventionist ideals, was the considerable extent of government involvement in both the development of technology in terms of foresight and its utilization.

Finally the paper set out the 'Specific government responses to the individual recommendations of the Trade and Industry Select Committee's report on optical fibre networks' (Command Paper 1994: p. 24). It made clear the government's commitment to continuing along its chosen course as set out in the 1991 White Paper (Command Paper 1994: p. 28). One specific recommendation of the Select

Committee of particular relevance here was: 'The government and Oftel, in association with the PTO's and cable companies, should seek to raise public awareness of the considerable public service benefits resulting from broadband networks' (Command Paper 1994: p. 29).

In response, the paper said:

> The government agrees that greater public awareness of the benefits of broadband communications would be valuable. Publication of this command paper is designed to help raise the profile of this issue, by setting out the government's vision of the importance for UK competitiveness of future developments in broadband. Greater public awareness should also help the companies involved be more responsive to consumer demand for a wide range of communications applications and services. Through dialogue and encouragement, the government will press all those involved to develop new and innovative services and to raise consumer awareness of the opportunities which are increasingly open to them.
>
> (Command Paper 1994: p. 29)

In responding to recommendations relating to 'public service benefits', the government made specific reference to 'competitiveness' and 'consumer demand' and did so in a manner clearly reflective of its overall ideological position which favoured the market over any public utilization of broadband technology. The government was committed to a programme of legislation, as has been seen above, that would see the technology develop along such lines.

When exposed to deeper analysis, there was a clear *policy of inaction* in which the lack of involvement was made explicit in the form of policy documents. To have been seen to have been part of the new developments would have given government a responsibility, a public responsibility for its utilization. This did not tally with official thinking. Instead the government sought, using separate pieces of legislation, to manipulate the situation so as to create a technological environment outside of the public sphere, an environment dominated by private interest, but private interest that would be subject to the conditions created by the dominant discourse in all its forms. The government, on the surface, simply left the developments to the market. But it was a market that it was very much in tune with, despite the fact that official policy in relation to convergent information technology was for the most part shaded from view.

Publicly, they would use it but would not promote its public utilization. As Jürgen Habermas has said: 'Of course, the political order remained unthreatened by the new processes which, as such, had no place in the existing framework, as long as the members of the old ruling stratum participated in them only as consumers' (Habermas 1979: 16). The quote at first glance accurately describes the situation, or more accurately, the perceived situation as it is being discussed here. The government, rather than having adopted new techniques and adapted accordingly, retained the established norms and patterns and was seen merely to consume the new technology within that established structure. It was very careful though to ensure that the new technology did not develop unchecked. The government contributed to its development via the use of strategic legislation – moulding it gently and almost invisibly into a compliant form within the confines of accepted practice. In doing so, it portrayed itself as vulnerable and as having to respond to technology in the same way as all other sectors of society.

Using 'dialogue' and 'encouragement', the government can be seen to have been successful in raising the public profile of broadband technologies and by 1994 there was widespread engagement with the new media and its predicted effects. A discourse on technology began to develop to a more significant extent from this point on as books, magazines, newspapers and television programmes began to embrace the topic.[9]

1994 was widely seen as an important year (Fidler 1997: 254). It represented a specific historical point according to *The Guardian*. On the 1st of January, developing a broad socio-cultural theme, it announced: 'the beginning of the 21st Century'. It hailed the dawning of the new media environment and its apparent sudden arrival in the public domain. 'In 1994, say those in the know, the world will fast forward into a high-tech future that will alter the way we think, talk and live' (*The Guardian*, 1 January 1994). In this high tech world, as the twentieth century moved towards its end, it was believed that traditional political power would become moribund and the 'popular' discourse also reflected this: 'Despite M.Ps' best intentions, parliamentary powers and those vested in the police are utterly eclipsed by ever advancing technology' (*The Guardian*, 16 April 1994). The dangers inherent within the technology for established political norms were thought to be so acute that it was further suggested:

> There is a world where governments have next to no power, and where children are equal, if not superior, to adults. It is a world of philosophers and bandits, of big business and science. It is a place where pornographers and Nazis walk freely, where criminals roam unchecked and where anarchy reigns. And it is the place where, perhaps, the future of the human race is being decided. It is real, it is here right now – and Britain is a growing part of it.
>
> (*The Guardian*, 30 April 1994)

Like Bachelard's poetic image, *new media* had captured the popular imagination and resonated in a way that caused a considerable degree of interest (Bachelard 1994: xvii).

Academic interest in the new technology was also on the increase during this period. In his article from 1995 entitled, 'Winners and Losers in the Information Age', Ian Angell, Professor of Information Systems at the London School of Economics, explored the political implications of new media and identified trends which he interpreted as causal factors in a process of fundamental change which went so far as to threaten the very legitimacy of the state. He was adamant that the move towards a new political order or what he called 'disorder' was due to developments in the area of telecommunications. Angell saw technology as being responsible for the decline of the state and the development of a kind of natural, or at least acceptable, political inequality. However, he did so without any wider assessment of the nature of technology, and in so doing abstracted technology from wider social and political processes. He cited phenomena which were interpreted as causal factors in a process of fundamental global change which, rather than facilitating the cohabitation of local discourses within a framework of liberal democratic states, had gone so far as to threaten the very legitimacy of the state. But why now, he asked.

> The answer is quite simple: a new order (which many will call disorder) is being forced upon an unsuspecting world by advances in telecommunications. The future is being born in the so-called superhighways. Very soon these electronic telecommunication networks, covering the world via cable and satellite, will enable everyone in the world to 'talk' to everyone else. We are entering a new elite cosmopolitan age. Global commerce will force through the construction of multi-media highways, and anyone bypassed by these highways faces ruin. Information technology, together with speedy

> international travel, is changing the whole nature of political governance and its relationship to commerce, and commerce itself.
>
> (Angell 1995: 10)

New media for Angell was new in the conventional linear sense in so far as it replaced what had gone before and had been itself caused by an inner technological thrust. There was no consideration of the complex political policy environment in which such developments occurred.

Pronouncements ranging from positive to negative, fatalistic to utopian, fantastical to the more realistic, were in abundance during this period as a discourse based on what Michael Heim has called the 'digital dialectic' began to proliferate (Heim, cited in Lunenfeld 1999). However, whether celebratory or sceptical in tone, each contribution to the discourse shared in common the predominantly liberal/market approach that was evident in the official statements above, and displayed a tendency to view technology as neutral and as developing within a largely unfettered free market. This was particularly the case in publications such as *Wired* magazine, which launched in the UK in 1995 with a cover adorned by Thomas Paine claiming: 'We have it in our power to begin the world over again.'

Equally enthusiastic about the potential of the new technologies was Bill Gates who wrote of 'new media' in his book, *The Road Ahead*:

> It will enhance our leisure time and enrich our culture by expanding the distribution of information. It will help relieve pressures on urban areas by enabling people to work from home or remote-site offices. It will relieve pressures on natural resources because increasing numbers of products will take the form of bits rather than manufactured goods. It will give us more control over our lives, enabling us to tailor our experiences and the products we use to our interests.
>
> (Gates 1996: 284)

In the US, similar discourses were proliferating:[10]

> The Magaziner report, approved in 1997 by the US government, is explicit in its recommendation on this matter: for e-commerce to spread at full capacity 'governments must adopt a non-regulatory, market-orientated approach to electronic commerce, one that facilitates the emergence of a transparent and

> predictable legal environment to support global business and commerce. Official decision makers must respect the unique nature of the medium and recognize that widespread competition and increased consumer choice should be the defining features of the new digital marketplace.'
>
> (Fischer 2006: 159, A Framework for Global Electronic Commerce, White House, 1 July 1997, http://www.pff.org/issuespubs/futureinsights/fi6.1globaleconomiccommerce.html [Accessed 10 Sept. 2014])

Similar claims were also expressed in the TV programme *Visions of Heaven and Hell*, the UK Channel 4 series from 1994 that aired the views of a number of commentators keen to predict the future.[11] As part of the emerging discourse that was forming, it further contributed to the finding of visibilities (Deleuze 1995). Films also contributed to this process: *Tron, The Lawnmower Man, War Games, Johnny Mnemonic, Strange Days, Hackers*, and culminating a few years later with *The Matrix*, all made a contribution to, as well as feeding off, the excitement of the moment.[12]

An interesting assessment of the way in which cinema operates at the level of discourse has been carried out by David A. Kirby. He has looked at the role of science advisors in Hollywood film. In doing so he stresses the importance of visual modes of representation as an effective means of demonstrating or bringing to life ideas that also exist in other forms, particularly in scientific papers and dialogues. This further relates to the idea of texts as thematically unstable and multimodal, rather than as singular and fixed within a medium. It occurs, he says, at the level of virtual witnessing. 'Fictional films, because they can allow large sections of the public to witness phenomena that are experienced as real can be considered as virtual witnessing technologies '(Kirby 2003). Kirby also claims that film has an important role to play in the process of knowledge formation. Indeed, at times, 'it forces consensus on the public by presenting a single vision of nature in a perceptually realistic structure' (Kirby 2003: 54–60).

Film specifically, but all visual forms more generally, can and do, according to this point of view, constitute statements within a discourse from the Foucauldian perspective of creating boundaries of possibility.[13] In the example under consideration here, they contributed significantly to the ordering and regimentation of a phenomenon that was, according to many, cited as inherently free of such restrictions. On this topic Ken Hillis has said: 'Film is central to organizing the cultural discourse around technology through its ability to popularize specific images of new technologies and to suggest a range of cultural attitudes that attend their introduction and social diffusion' (Hillis, www.unc.edu/~khillis/techrep.html [Accessed 10 Sept. 2014]). Political, academic, and popular statements fused to form an environment wherein novelty resonated 'for a mind that is foreign to the process of its creation' (Bachelard 1994: xvii).

What all of this shows is a clear coherence and resonance of statements in a range of forms. It shows how politics, and responses to it, take place in the same space as system and lifeworld morph into one another. This system/lifeworld dualism appeared to be in evidence as government sought explicitly to position themselves on the outside of any causal chain of events, preferring instead to present themselves as subject to technological effects. But this was not the case: they were neither inside or outside but remained and still remain an important player in a network of statements and practices that combine to form technological discourses that are inhabited, constructed and responded to simultaneously.

Approaching discourse as mediated in multiple forms allows us to see the continued importance of the political realm without claiming it as a necessarily determining force. A lifeworld must always already be a political realm – a site of struggle – not a neutral venue ripe for colonization or a site of potential resistance. Life takes place *in* the system. The development of *new media* must then be seen to have taken place not within a vacuum but within an environment of political structures, institutions, established patterns, restricted or encouraged accordingly by specific pieces of legislation and supported by a dominant discourse. Specific statements on technology at the political level are always related to statements at other levels within what might be commonly be described as a cultural or media realm. They operate at the intersection of political discourses and their associated cultural forms that are manifest in film, TV or indeed music and other art forms. Together they make up the technology agenda. It is not a random process but one that operates in accordance with an agenda. Yet, as we have seen, that agenda need not always be coherent in terms of themes or stylistic approaches. Dispersed statements across time and space came together in 1994/95 to form a discourse, a crescendo or a Zeitgeist. It was a discourse that took a multiplicity of forms.

Discourse forms as links between popular statements and the official policy statements connect. The discourse *on* technology examined here sought to separate itself *from* technology. But this can never be the case. *New media* did not spring forth as if from nowhere, as is often assumed, but neither can its existence be attributed to a standard cause-and-effect chain of events. The above has shown how a complex contrivance of multiple and interrelated factors combined in such a way as to create a moment (Lefebvre 2004: x). Neither did it fade away entirely. Rather it must be understood as a process that was and is always in-formation and subject to varying intensities. At the end of the 1990s it did not cease to be but it did plateau and become assimilated for a period before once more intensifying. Hence *new media* does not name an age that can be assigned a clear linear definition, but a technological phenomenon that is always in process relative to the conditions of its existence and where space emerged as a space relative to real-world political concerns.

Such complex networks or discourse environments, it is argued, work like musical refrains, and they do so in a way where sound is more is important in itself and not just as a metaphor. As Brandon LaBelle has said:

> [F]or in traveling away from itself, sound is picked up elsewhere, overheard, carried forward, or brought back, through memories and recordings, to enliven the making of social space. Acoustic territories are then specific while being multiple, cut with flows and rhythms, vibrations and echoes, all of which form a sonic discourse that is equally fever-ish, energetic, and participatory. Sound is shared property onto which many claims are made, over time, and which demand associative and relational understanding.
>
> (LaBelle 2010: xxiv)

In order to make sense of this complex sonic discourse it is useful to refer also to Henri Lefebvre's *Rhythmanalysis*. Lefebvre set out a reconfigured kind of political analysis and one that is significant in relation to this book. He said:

> Social times disclose diverse, contradictory possibilities: delays and early arrivals, reappearances (repetitions) of an (apparently) rich past, and revolutions that brusquely introduce a new *content* and sometimes change

> the form of society. Historical times slow down or speed up, advance or regress, look forward or backward. According to what criteria? According to representations and political decisions, but also according to the historian who puts them into perspective. Objectively, for there to be *change*, a social group, a class or a caste must intervene by imprinting a rhythm on an era, be it through force or in an insinuating manner. In the course of a *crisis*, in a critical situation, a group must designate itself as an *innovator* or *producer of meaning*. And its acts must inscribe themselves on *reality*.
>
> (Lefebvre 2004: 14)

This relates directly to the case study below where reappearances occurred in locations other than their original revolutionary inception, to herald new times amid crisis. Only they were not so much 'appearances' as echoes.

In *A Thousand Plateaus*, Gilles Deleuze and Félix Guattari also show how music contains certain territorial characteristics that are immanent and part of wider discourse and a political strategy of resistance (Deleuze and Guattari 1987: 342–86). 'In Deleuze and Guattari's terminology, bird calls are largely milieu components, fixed to specific functions, whereas birdsongs are territorial elements, sonic components of milieus that have been unfixed and reconfigured in a more autonomous fashion' (Bogue, cited in Khalfa 1999: 114). Sonic milieu components consist of sounds that perform a specific function – marking out the availability of food, warning of danger, etc. They are the noises our world makes. The refrain, on the other hand, organizes sounds together, bringing order to chaos. It can be but is not always sonic. It consists of organized rhythms and patterns; daily routines and habits, the justification for which has long been forgotten. It marks out the territories that we recognize as our own or as belonging to others. It creates a sense of familiarity and belonging yet can also bring about alienation and a sense of detachment. It is our local neighbourhood, our political system and our state. Music, according to Deleuze and Guattari, can also deterritorialize this ordered terrain – disrupt the rhythm, creating a momentary energy field that distorts the dominant order of things. Of course, not all music does this. So it is up to us to develop arguments drawing on the theoretical concepts mentioned thus far to justify classifying a particular musical moment as disruptive and resistant, and this will be done in Chapter 3.

The intention here is to develop a spatialized sense of music and its political significance based on the concept of the refrain, but also drawing on Heidegger, Foucault, Lefebvre and Winner.[14] What is important is to assert music/art as resistance strategies not in terms of dialectical oppositions but as positive and legitimate acts of rebellion that at their best serve to unsettle the territorial familiarity and the dominant political order, and not just operate as distractions from more so-called serious issues, or as activities of aesthetic escapism. 'The revolutionary spirit, born of total negation, instinctively feels that besides refusal, there was also in art a tendency to acquiescence; that there was a risk of contemplation counterbalancing action and beauty counteracting injustice, and that in certain cases, beauty itself was a form of injustice from which there was no appeal' (Camus 1965: 223–4).

In this context, music is often seen as something transcendent; a thing of beauty created almost without thinking. In this sense it is too often regarded as something inexplicable that has the ability to affect in a manner beyond explanation. To counter this, music and sound must be posited as immanent rather than transcendent: immanent not *to* something but only to itself. It is significant in both theory and practice, so much so that listening to and talking about music, and not just making it, are regarded here as creative practices at the political level: 'talk of creativity is not bought at the cost of further creativity on the part of the person who talks about it' (Diffey 2004: 93).

This sentiment is echoed here:

> Rather than protecting music as a sublimely meaningless activity that has managed to escape social signification, I insist on treating it as a medium that participates in social formation by influencing the ways we perceive our feelings, our bodies, our desires, our very subjectivities – even if it does so surreptitiously, without most of us knowing how. It is too important a cultural force to be shrouded by mystified notions of romantic transcendence.
>
> (McLary 1994: 205–35)

As a 'cultural force' its political and economic significance has not gone unnoticed by the political establishment. Hansard recorded on the 13 March 2007 the Conservative MP, Mr Edward Vaizey as having said:

> It is interesting that tonight in New York Malcom McLaren will promote the British music industry under the DTI's auspices. Our cultural institutions need a one-stop shop, as recommended by the pamphlet, whereby if they have a proposal and want to do something abroad they can walk through one door and have the expertise of all four departments available.
>
> (www.publications.parliament.uk/pa/cm200607/cmhansrd/cm070313/halltext/70 [Accessed 10 Sept. 2014])

Music in this instance can be seen as a refrain, enframed as standing reserve, and ready for export. Exported that is, as part of a broader refrain that serves to echo and promote Britain's cultural self-confidence (aka imperialism) – referred to in the Demos pamphlet mentioned above as 'cultural diplomacy' (http://www.demos.co.uk/projects/culturaldiplomacy/overview [Accessed 10 Sept. 2014]).

So music can operate, in what Brandon LaBelle calls an acoustic territory, as both a commodified product or as an aspect of deterritorialization. Either way, it is organized as a refrain with a degree of autonomy that means it always belongs to a political economy in one way or another.

Music here is positioned as a relative aspect of a chaotic media environment. Such media environments are always specific discursive formations that give bounds to the world in the Heideggerian sense. The ordered formation of an environment might become either enshrined or unsettled through music. Hence music, like the birdsong mentioned above, can be regarded as inherently political – always either territorializing or deterritorializing. Music, as it will be discussed below, is about politics and protest as actualized events on a virtual plane, or rather about a politics *of* protest, and of rebellion relative to the environment in which it occurs. But it is also about creativity and the way in which theory and practice, or the virtual and the actual, combine within the context of the 'productive/creative' process.

Locating the creative product in such close proximity to critical analysis raises important issues in relation to the creative process itself. Questioning the tendency to privilege the creative over the analytic, T.J. Diffey has said:

> At the root of my prejudice is the idea that a work of art is a finer thing than talk about the creation of art ... When this prejudice is confessed, we can see what is wrong with it. Spelled out, it holds that imagination

> is superior to intellect. Therefore art and poetry [add music for the purposes of this chapter], as works of imagination, are superior to intellectual analysis. Talk of creativity is intellectual analysis of creativity and therefore inferior to creativity itself. Better to be creative than to talk about it.
>
> (Diffey 2004: 92)[15]

Is it better to produce art than to talk about it? To answer yes infers a mystery of creativity or transcendence, or the presence of a genius, a human creative source. This seems to be particularly true for music, an arena where intellectual debate is often met with derision – after all, it's all about *soul* and *feeling*. In challenging this kind of approach, it is important to acknowledge the 'New Musicology' of Susan McClary who endeavoured to move away from the practice of treating music as an abstract phenomenon remote from political signification. This practice is similarly resisted here where music is assigned a large degree of political significance.

Doing so raises the question of whether music is immanent *to* anything within time and space. The sound of the times, the obsequious sound of the sixties or seventies or whatever decade is enframed, re-packaged and sold as a nostalgia trip? The answer is a categorical no. 'Deleuze calls the plane of immanence (or consistency) "the image of thought." It is not immanent to something, but only to itself. "Whenever immanence is interpreted as immanent to something, we can be sure that this something reintroduces the transcendent"' (http://www.christianhubert.com/hypertext/plane_of_immanence.html [Accessed 10 Sept. 2014]).

Music then is similarly immanent only to itself – knowable within the conditions of its existence and subject to perpetual critique therein.

Protest or politically engaged music of the kind discussed below is immanent, not to some specific time, place, or disenfranchised group, only to be repeated as a packaged memory of a bygone era. To categorize it thus would serve to reintroduce some kind of transcendent element – some unknowable quality. That is not to say there are no economic, geographical, or temporal characteristics to it, simply that it sits along an axis of 'perpetual critique' as it ebbs in and out of time with the conditions of its existence.

Notes

1 Digital Economy Act 2010, http://www.legislation.gov.uk/ukpga/2010/24/content [Accessed 8 June 2014]; Digital Economy Funding Link: http://www.ahrc.ac.uk/FundingOpportunities/Pages/DigitalEconomyProgramme.aspx [Accessed 8 June 2014]; Digital Humanities Links: http://www.ucl.ac.uk/dh/ [Accessed 8 June 2014]; http://digitalhumanities.org/dhq/ [Accessed 8 June 2014]; Wired Link: http://www.guardian.co.uk/media/2009/mar/23/wired-magazine-media-technology [Accessed 8 June 2014]; BBC Superpower Link: http://www.bbc.co.uk/worldservice/specialreports/superpower.shtml [Accessed 8 June 2014]; Virtual Futures Link: http://virtualfutures.co.uk/ [Accessed 8 June 2014].

2 As part of this renewed interest, earlier claims relating to *new media* and its potential impact are being revisited in order to judge whether the term is one that still makes sense. http://www.reportr.net/2009/02/12/why-new-media-is-a-generational-term/ [Accessed 8 June 2014]. In 2006, BBC re named its New Media Dept, Future Media & Technology. See http://www.guardian.co.uk/media/2002/dec/11/newmedia.bbc [Accessed 8 June 2014].

3 Newness here is explored as a concept in relation to the work of Gaston Bachelard:

> To say that the poetic image [**discourse on New Media Technology**] is independent of causality is to make a rather serious statement. But the causes cited by psychologists and psychoanalysts [**Technological determinists, Economic determinists or those who resort to instrumental and representational thinking**] can never really explain the wholly unexpected nature of the new image, any more than they can explain the attraction it holds for a mind that is foreign to the process of its creation.
>
> (Bachelard 1994: xvii; bold not in original)

Like Bachelard's poetic image, the phenomenon being referred to as *new media* can be explored as being without cause. *New media* as phenomena were not caused by their predecessor *old media* nor indeed did they supersede them. In order to dispel simple cause-and-effect explanations in relation to *new media*, the dispersed nature of its coming into being must be examined with no single aspect being assigned any greater or lesser weight. The aim here is to show how the constituent parts were always in existence and to highlight the chaotic nature out of which patterns emerged whose origin need never be established. Equally the technology associated with *new media* did not have a simple political effect because we can identify political statements that contradict that claim and that show how formal political structures were complicit in the creation of the phenomena of dispersed factors that in combination, operated as an agenda that contributed to the process of its coming into being. Attempts to fully grasp what was happening during the period under investigation here have too often failed to engage with this idea and relied instead on traditional notions of cause and effect, and to a large extent still do. Failure to properly understand what happened as part of the new wave of technological enthusiasm in the mid-1990s means it will be impossible to understand what is happening now. It is important to ask what has changed between now and then. When the discourse returns, is it still new? Was it ever new or does such a notion always imply a causal thrust? In order to answer these questions the recursive nature of *new media* needs to be understood in a way that does not rely on concepts of new and old.

4 *Past* (Californian Ideology, available at: http://www.hrc.wmin.ac.uk/theory-californianideology [Accessed 8 June 2014]; html *Wired* UK Issue 1 1995 with the valorization of Marshall McLuhan and Thomas Paine available at: http://yoz.com/wired/1.01/index.html [Accessed 8 June 2014]), *present* (as seen in documentation available in archives at any given moment), and *future* (the largely predictive nature of technological discourse is readily identifiable) combined to create the discourse under investigation here.

5 The use of the word version requires some clarification. It is used to describe the particular manifestation of a phenomenon that could – and still may yet – have taken innumerable different forms. As such, it describes something that differs from its predecessor without having replaced it or having been caused by it.

6 For a more detailed discussion, see http://www.csudh.edu/dearhabermas/habfouc02.htm [Accessed 8 June 2014]. An alternative mode of analysis is explored in Kennedy (2011).

7 See Golding (2011).

8 Information Technology Advisory Panel, *Making a Business of Information: A Survey of New Opportunities* (London: The Cabinet Office, 1983). The Hunt Report, *Inquiry into Cable Expansion and Broadcasting Policy*, Home Office Inquiry (London: The Stationery Office, 1982).

9 This is a process that relates to Virilio's work on global and local time and the imposition through a kind of ubiquitous advertising of a new mode of being as digital convergence displaces its analogue predecessor (Virilio 2005: 107–14).

10 http://www.oecd.org/regreform/2506672.pdf [Accessed 8 June 2014]; see also Walter G. Bolter, James W. McConnaughey, and Fred J. Kelsey, *Telecommunications Policy for the 1990s and Beyond* (New York: M.E. Sharpe, 1990).

11 Many of themes covered in *Visions of Heaven and Hell* (1994) (http://www.imdb.com/title/tt0298212/ [Accessed 8 June 2014]) return as echoes in *Visions of the*

Future (2007) (http://video.google.com/videoplay?docid=80785608139596344) [Accessed 8 June 2014] as the discourse reconfigures itself in terms of both form and content.

12 This process continued into the twenty-first century and while the films cannot be fully explored here, their role in the discourse process should be taken very seriously. For a fuller discussion, see Adil and Kennedy (2009: 219–30). They write:

> Film and its visualisation of other technologies can and should be seen as *operational* within discourse, existing not solely in the symbolic realm but bound up in the world materially. But what is a discourse? What is it that film is a part of? For Michel Foucault a discourse is an amalgamation of statements that exist within a specifically prescriptive modality that allows them to have a material impact. The films examined here are statements within a discourse that form a relationship with the 'real'. It is not an empty reality composed of representations in the manner put forward by Slavoj Žižek or Jean Baudrillard. Films do more than simply reflect or set up a model for imposing vacuity wherein the real world is reduced to a screenplay – rather they make varying and specific contributions to the nature of that reality.

13 See Kittler (1999).

14 Music can be seen as a statement that coheres with other statements to create a discourse (Foucault), the intention of which is to enframe (Heidegger) within a specific space. A space that has been created rather than being the consequence of other more dominant forces (Lefebvre). An example of this is the Harlem River Drive project discussed by Langdon Winner in 'Do Artifacts Have Politics?', *Daedalus*, 109(1) (1980): 121–36.

15 See also Diffey (2004: 91–102).

3

Echostate[1]

Visual logic, as discussed above, is prefaced upon the imposition of order, as has been established by a historical evaluation of its coming to prominence. This proceeded from Plato, through Aristotle, and Nietzsche as key reference points, and up to McLuhan via Francis Bacon, and Don Ihde, to explore a new acoustic, or sonic mode of thinking. What has become clear is that visual logic is not well suited to chaotic situations. Contemporary digital space constitutes just such a situation.

Visual representations have a spatial bias and serve to fix phenomena and categorize them in an ontic fashion where similarity becomes a means of organizing difference. From such a perspective as this, semiotics attains a dominant position. Acoustic or sonic phenomena, on the other hand, have more of a temporal bias where scale cannot be discerned in a unified fashion. Like Lynch's temporal city, the digital environment is abuzz with vibrations: not a taxonomy of visual similarity but a cacophony of sonic difference. Relationships between images and their referents in visual space are spatially and temporally fixed. Musical notes, on the other hand, are meaningless on their own but cohere in acoustic space temporally and temporarily as affective forces. This quantum, vibrating, 'fading into permanence' as Jean-Luc Nancy calls it, seems like a better way of coping with the complexity of digital environments.

Aesthetic concerns have been central in developing the challenge to visual logic and its inability to account for the complexity of such environments. Where Nicolas Schoffer wanted sonic arts to take their cue from the visual, Joseph Nechvatal has recently reversed this position to invoke the sonic in the visual with noise as the central theme. A sonic economy follows neither of these paths. It proposes instead an approach based not on space (visual) *or* time (sonic) but the relativity of space–time. This is because the digital environment that we currently inhabit is analogous to a chaotic curiosity shop that, and like the cities we inhabit for 'real', must be understood as temporal as well as spatial (Lynch 1960). Bewilderment as well as scale prevents the space being perceived in its entirety as was identified by Foucault via his engagement with Raymond Roussel. The analogic motif of the curiosity shop derives from Jacques Rancière who, remarking on the importance of embracing new kinds of aesthetic thinking in terms of understanding wider social phenomena, said:

> Now this multi-temporality also means a permeability of the boundaries of art. Being a matter of art turns out to be a kind of metamorphic status. The works of the past may fall asleep and cease to be artworks, they may be awakened and take on a new life in various ways. They make thereby for a continuum of metamorphic forms. According to the same logic, common objects may cross the border and enter the realm of artistic combination. They can do so all the more easily in that the artistic and the historic are now linked together, such that each object can be withdrawn from its condition of common use and viewed as a poetic body wearing the traces of its history. In this way the argument of the 'end of art' can be overturned. In the year that Hegel died, Balzac published his novel *La Peau de chagrin*. At the beginning of the novel, the hero Raphael enters the show-rooms of a large curiosity shop where old statues and paintings are mingled with old-fashioned furniture, gadgets and household goods. There, Balzac writes, 'this ocean of furnishings, inventions, works of art and relics made for him an endless poem'. The paraphernalia of the shop is also a medley of objects and ages, of artworks and accessories. Each of these objects is like a fossil, wearing on its body the history of an era or a civilization. A little further on, Balzac remarks that the great poet of the new age is not a poet as we understand the term: it is not Byron but Cuvier, the naturalist who could reconstitute forests out of petrified traces and races of giants out of scattered bones.
>
> (Rancière 2002: 1)[2]

This chapter proposes the concept *echostate* as a poetic strategy for making sense of the 'multi-temporality' of digital space. *Echostate* is a term related to Michel Foucault's category 'statement', in combination with Gaston Bachelard's use of 'reverberation' and 'echo' in his introduction to *The Poetics of Space*, and expands their scope by examining how statements are echoed and amplified across a range of media that now exist in multiple simultaneous forms – repeatable but not always identical in as much as they can take on numerous different digitally enabled forms.[3]

Statements are more than abstract signs. They are operational performances that cohere with objects to create a kind of materiality that is repeatable and melodic, rhythmic and harmonious. They are never to be cited themselves as examples for fear of rendering them static and empty. Instead they modulate incessantly in a dispersed fashion that makes them difficult to tie down and to have representational meaning assigned to them. This is why they are so significant and powerful. They very definitely exist but not always in the way that we expect them to.

In relation to this, Michel Foucault said:

> We will call *statement* the modality of existence proper to that group of signs: a modality that allows it to be something more than a series of traces, something more than a succession of marks on a substance, something more than a mere object[4] made by a human being; a modality that allows it to be in relation with a domain of objects, to prescribe a definite position to any possible subject, to be situated among other verbal performances, and to be endowed with a repeatable materiality.
>
> (Foucault 2004a: 120)

It is the being 'in relation with' that is important here. Statements move through objective space, by which is meant digital space as unified, real and devoid of liminal thresholds between real and virtual. They form bonds with other statements to create conditions of possibility or frameworks, a kind of coherent space, where actualities arise from the virtual ground within which individuals become accustomed or attuned as to how to operate.

In so far as the coherence of statements operate to form discourses and conditions of possibility, they are witnessed as echoes identifiable in what Nigel Thrift calls 'practice' (as detailed below) and which, as the third tenet of non-representational theory, serve to sustain a veneer of stability while simultaneously serving as a means of identifying the changing state of things.

These systems of formation, despite appearing stable in so far as they represent a common-sense reality, are in a perpetual state of uncertainty. Both statements and discourses move and return as echoes – sometimes clear and intact, recognizable and coherent, but always with the possibility that they will come back altered or rearranged, or less recognizable, allowing a new or different common sense to prevail. They operate as networks that are never static, but always operational. For as Bruno Latour (2005) has pointed out, to become static is to cause the network to dissolve.

Like Bachelard's poetic image, discussed in Chapter 2, they are formations that are not determined in any causal linear sense, by the arrangements that precede them.[5] They are not objects, not substitutes for objects, but resonances that are to be understood on their own terms as real and material but with an as yet uncertain form.[6] 'The poetic image is not subject to an inner thrust. It is not an echo of the past. On the contrary: through the brilliance of an image, the distant past resounds with echoes, and it is hard to know at what depth these echoes will reverberate and die away (Bachelard 1994: xvi).

Echoes persist, forming alliances and environments out of which new poetic forces emerge – relative to but not caused by their predecessor. This instance, this new moment serves to amplify the fading echo rather than making itself reliant on its historical force. Hence echoes are discontinuous for Bachelard. Bachelard goes on to make the dialectical distinction between soul and mind with the poetic image, or the new creative moment, residing in the former before it is formalized as representation in the latter. So artists and poets, as well as those who engage with their work, dwell in the nonrepresentational realm of the soul where affect reigns. The dichotomy of mind and soul like McLuhan's environment and anti environment (McLuhan and McLuhan 1988) seem problematic in so far as the artist is situated as dialectically opposed to the so-called real world. This dialectical opposition means that whether by stealth or intention, the polarized dualism, as discussed above, returns. Indeed, Bachelard stresses this when he says: 'Forces are manifested in poems that do not pass through the circuits of knowledge' (Bachelard 1994: xxi).

This point needs to be thought through very carefully: the usefulness (if indeed they need to be useful) of creative pursuits is surely to open up, as Foucault has done, avenues of exploration into social and political formations that are themselves not indifferent or immune to artistic expression. So established circuits of knowledge only tell part of any story. But the point is not to separate them off from other different forms of knowledge, or to create hierarchies, but to recognize the constant interplay between them in a unitary but relative space. This was demonstrated in Chapter 2 in terms of Foucault's exposition of Raymond Roussel's work.

Like Bachelard's poetic image, the formation of a discourse is equally not caused by or linked to the past in a linear fashion but is a creative process or 'moment' as a discrete unit of time that might appropriate elements of the past – re-animating, re-mixing them or amplifying a fading echo that had been dying away or that had been silenced (almost but never quite) in an archive. When coherent, discourses (as refrains) have an amplitude and a harmony that is significant enough to mark them as recognizable practices, their constituent statements can dissipate and fragment at any time, or when the 'moment' is right.[7] Identifying such formations, or 'moments' of dispersion is what Foucault does and it is a process that can be further attuned by the introduction of the sonic so that the dispersed invisibility of arrangements, their stealth, does not render them inaccessible and therefore beyond critique. When obscured by shadows, statements may still be heard, and for this reason, the emphasis on the solely visual must be challenged.

This privileging of the visual realm produces a particular kind of knowledge; a representational kind that names and fixes and renders concepts and phenomena stationary. And it is a system of thought that works, it is reliable and we have grown accustomed to it. But it is also limited. It leaves out much of the important information or is simply incapable of mediating the intensity of feeling that may be present yet difficult to communicate. As Marco Polo explains to Kublai Khan, when describing a scene from his travels in Italo Calvino's *Invisible Cities*:

> I could tell you how many steps make up the streets rising like stairways, and the degree of the arcades' curves, and what kind of zinc scales cover the roofs; but I already know this would be the same as telling you nothing. The City does not consist of this, but of relationships between the measurements of its space and the events of its past . . .
>
> (Calvino 1978: 9)

To avoid 'telling you nothing' and to communicate the multiplicity of the lived moment that resounds with echoes, may require more than simple representational devices then. The space between measurements as it is considered in this case is not an *in-between* as in between point *a* and *b* or *a* to *z*, but a thoroughfare or a medium that is alive and relational, through which sound vibrates and echoes. It is a space where possibilities abound and where multidimensional elements or statements resound, compiling, combining, cohering as practices or discourses. We are aided in the task of understanding this by Heidegger in so far as he offers the possibility of a challenge to the dominance of Platonic logic that came to dominate Western philosophy and science, by proposing a re-focused critique of Being or a 'fundamental ontology'. Such a proposal takes us into the realm of the invisible or the not-so-visible world of shadows where contemplation rather than a gazing upon, prevail.

Such contemplation may paradoxically be best done in silence. Sara Maitland has suggested that silence helps us to not only listen better, but also to see better – to be more attentive and detailed in our observations (Maitland 2009: 157). But there are different kinds of silence: the silence of the desert and the silence of the forest.

> The Silence of the forest is about secrets, about things that are hidden. Most of the terrains of silence – deserts, mountains, oceans, islands, moorland – have austere but wide views. They are landscapes that can be appalling in their openness, but at least you can see what is coming. The wide sky is bright above you, the clouds give you warnings of approaching weather and the land sweeps away into the distance. But the silence of the forest hides things; it does not open them out but closes them off. Trees hide the sunshine; and life goes on under the trees, in the thickets and tanglewood. Forests are full of surprises. It is not strange that the fairy stories that come out of the forest are stories about hidden identities, both good and bad.
>
> (Maitland 2009: 176)

These hidden identities are present, Maitland says, in Cinderella, and in Snow White, where the true nature of the character is masked, and most explicitly in Rumplestiltskin, where the silencing of the name, the mystery of the character's true identity is what gives him his power: to know his name is to take away that power. This silencing of identity will be seen to be present in Chapter 5 where the name of the band The Special AKA, will be shown to be indicative of a struggle for control and identity, and also in the case of the Detroit techno collective Underground Resistance, whose members' true identity is fiercely guarded.

Junichiro Tanizaki also takes up this point in his (2001) work, *In Praise of Shadows*, in which he highlights, in a manner comparable to Heidegger, the Western obsession with light and seeing (a process of getting nature 'in hand' that Heidegger calls enframing) that took Eastern science and technology in a direction that was not its own. 'Imagine,' he asks:

> if we in the Orient had developed our own science. Suppose for instance that we had developed our own physics and chemistry: would not the techniques and industries based on them have taken a different form, would not our myriads of everyday gadgets, our medicines, the products of our industrial art – would they not have suited our national temper better than they do? In fact our conception of physics itself, and even the principles of chemistry, would probably differ from that of Westerners; and the facts we are now taught concerning the nature and function of light, electricity, and atoms might well have presented themselves in different form.
>
> (Tanizaki 2010: 14)

Even fundamental scientific principles then are subject to the fluid interplay of statements within discourses, to the arbitrary formation and amplification of ways of thinking that are heard above all others and seen in practices that come to dominate at the expense of dissonant expressions struggling to establish themselves as statements.

The emphasis on light and vision and observation impacted on all aspects of scientific, artistic and cultural life, according to Tanizaki. It even imposed itself as statement, or an echo of a statement (echostate) where contrary conditions of possibility existed but failed to take or keep hold. It may be that the power of light brings only a particular kind of knowledge.

Tanizaki goes on to say: 'It has been said of Japanese food that it is a cuisine to be looked at. I would go further and say that it is to be meditated upon, a kind of silent music evoked by the combination of lacquerware and the light of the candle flickering in the dark' (Tanizaki 2010: 26).

There is much to be gained from entering this shadowy realm, and listening to the silent music of contemplation. If embraced, it can prompt a revealing. As Heidegger says in his essay, *The Question Concerning Technology*:

> Always the unconcealment of that which is goes upon a way of revealing. Always the destining of revealing holds complete sway over man. But that destining is never a fate that compels. For man becomes truly free only insofar as he belongs to the realm of destining and so becomes one who listens and hears, and not one who is simply constrained to obey.
>
> (Heidegger 1977: 25)

For now, however, the question that needs to be asked is, how does this relate to digital space? Does it hide its true nature behind its vast incomprehensible expanse, or disguise itself as sensible phenomena without ever revealing its inaccessibly small constituent elements? These questions are different in kind and relate to the macroscopic cosmos and the microscopic levels accessible only through a kind of looking that embraces silence. The importance of the auditory realm as a means of enhancing looking is central here as an element of a sonic economy that operates in acoustic rather than visual space. In the same way that the forest closes down and keeps secrets in its dark shadows, digital space is similarly shaded. These are the shadows that we need to penetrate and in so doing we invoke Foucaut's Rousselian journeys of encounter and uncovering without revealing.

This journeying requires a particular kind of looking that has an aural dimension of movement and time. So with the emphasis on sound and vibration and motion we enter an immersive universe that is multi-sensual in and beyond three-dimensional space. But the silence that is so cherished by Maitland may not always be what it seems: vibrations persist at the cosmological and quantum levels, meaning there is always noise.

But what is noise and how does it relate to the task at hand here? Greg Hainge gives a comprehensive account in his study, *Noise Matters: Towards an Ontology of Noise* (2013), and it is an account that merits detailed consideration. Reflecting on the long-held assumption that noise is a dissonant opposite of music, Hainge points to the historical tendency of the latter to join forces with the former, as noise has itself become a musical genre. He notes: '[I]f noise can become what it is not, what exactly is it?' (Hainge 2013: 2).

Hainge explores the question what is noise from a number of points of view. From the point of view of physics, he says, it is a collection of sound waves of different frequency that do not harmonize. Or, from another point of view, it is a sensation that is disagreeable or uncomfortable: here a subjective measure is added to the mathematical definition of the physicist. This introduction of the subjective once again raises the polarized dualism discussed above and its reconstituted proposition from Quentin Meillassoux, that proposes the existence of a mathematical real, the digital, and a phenomenological real.

Another account of noise is based on principles of signal interference as set out by Shannon and Weaver in their study, *The Mathematical Theory of Communication* (1949). Noise for them is characterized by the addition of things to a signal that were not intended at source. The resulting distortion can, they say, affect any signal in any medium – radio, TV, telegraph. So, for them, noise is not exclusively sonic, a point that is also emphasized by Joseph Nechvatal, in his *Immersion into Noise* (2011). The important question for Shannon and Weaver was how the noise affected the original signal and its intention, and ultimately how the interference could be minimized or eliminated.

Responding to this position, Hainge points further to the increase in noise in the twentieth century and to the prevailing notion that it is something to be reduced or eliminated. However, the drive towards minimizing noise leads ironically, he says, to greater noise. For this reason, we need to embrace a redefined sense of noise, and to celebrate its ineluctable character – not to avoid it but to engage with it. But attending to noise is not easy as its definition is largely subjective and unstable. In trying to bring some kind of order to the issue of noise, Hainge notes the contributions of Bart Kosko in *Noise* (2006) and Garret Keizer in *The Unwanted Sound of Everything We Want: A Book about Noise* (2010). Both highlight the double nature of noise as something affective (of the heart) and as something that can and must be defined (the head). Once more here we seem to encounter the dreaded dualism. He also refers to Hillel Schwartz and his exploration of all the meanings that have been attributed to noise. Schwartz, however, ultimately fails to arrive at a consensus and instead seems to celebrate the chaotic 'self-contradictory' nature of noise.

This is not enough for Hainge whose stated task is to tie down what it means to talk about noise and to show why such an aim matters (Hainge 2013: 7).

Hainge wants to bring the previous definitions together to demonstrate consistency because of the increased interest in noise as a phenomenon. To do so is to prevent noise being reduced to a meaningless cacophony of merely subjective points of view: 'This, however, is just how noise has been treated, as it has been used to apply to everything and nothing at the same time, subject to a whole host of mutually contradictory definitions and usages, its apparently ineffable nature the result of divergent agendas rather than something proper to noise itself' (Hainge 2013: 8). He wants to form an operational taxonomy of noise – or to establish noise ontologically – prefaced by the contention that all ontology is necessarily relational.[8] The idea of noise as something that occurs in relative space is an important step in relation to the task at hand here: to develop a sonic methodology capable of accounting for the perceived chaos of contemporary digital space.

Noise has been predominantly, though not exclusively, regarded as auditory and aleatory, disagreeable and uncomfortable – especially for Keizer who analysed decibel levels and physical harm – noise thus entering the realm of power and domination.[9] From a slightly different angle comes George Prochnik who, though in agreement that noise is disagreeable, approaches it from the point of view, not of domination but of resistance. It is, he believes, the subjugated masses' refusal to be silenced. Hence it is regarded as oppositional (again dialectical) – noise is regarded as a by-product of another action, in this case, protest. As in information theory, the noise attaches itself to a signal and distorts its source intention. In this case, however, noise may not enter the signal from outside but gather force and amplitude from within the communication channel itself. For George Prochnik (2011), noise operates in a closed environment, or in absolute space. For him, everything vibrates at its own frequency when acted upon by an energy not from outside itself but from inside the system. But in essence, he claims, everything wants to be still and silent. Noise then is the result of things being disturbed and ultimately they will try to return to rest. It is opposition that stirs the unrest and causes the noise. Noise from this perspective is the result of something rather than something in itself.

Noise in such a context is credited with a disruptive political function. This is also the approach adopted by Jacques Attali. For him, it is a dialectical force that, once its disruptive function is complete, is absorbed back into the dominant code. For Attali, noise is doubled, as the sound of opposition, but an opposition that is prefigured by a musical opposition – so here music is itself noise and not its opposite. Music from this perspective is an avant-garde, an early indicator of wider social change – it jars and disrupts before being reintegrated.[10] It can, Attali believes, provide a means of predicting what is to come, as will be discussed in Chapter 4. But, for Hainge, this resistance is not necessarily political, in so far as it is regarded as standing as the dialectical opposite of a prevailing political force. Rather, it as an expressive force, it resists in another way, a way that foregrounds process rather than revolution (Hainge 2013: 16).

The dialectical approach is regarded here as overly simplistic.[11] What is required is an approach that is properly grounded philosophically, and Hainge makes a valuable contribution to this while maintaining an awareness of the related problem that to finally bring noise, and for that matter chaos, under control is to render it static and to strip it of the qualities that make it worth engaging with in the first place.

The anti-ocular turn (Janus 2011) has seen a growing degree of interest in sound/sonic/noise and the auditory more generally, including contributions from Nancy, Ihde, Dyson, Serres, etc. But Hainge wants to pursue a more general expanded definition of noise, one that includes the visual, one that embraces and advocates the significance of vibration but that also goes beyond it: to posit noise as the immanent material in-between of the virtual and the actual as event. 'Noise, rather, will be figured here as the trace and index of a relation that itself speaks of ontology' (Hainge 2013: 13). The use of the word 'trace' here is interesting in terms of its relationship to deconstructionism and in turn to existentialism, and it prompts a slight diversion. Hainge offers up a mix of Derrida, Deleuze and Whitehead in his attempt to use noise as an organizing principle, as well as a metaphor for wider ontological questions. As such, he is a monist, and his model does not require that the Cartesian dualism be reinstated, as is the case with the speculative realists. His is an approach that is largely echoed here, but with one small reservation, or maybe a question. Does the assertion that noise is a trace, reposition the human as a privileged site of activity? If it does, then the work of Latour and the Actor Network Theorists, as well as the object-orientated philosophers, that has done much to recognize the material significance of factors beyond the human, is lost, and that is not the intention here. Rather the intention is to develop a model that embraces sonic materiality in such a way that positions humans as actualized events, arising from a virtual plane. Once actualized, humans take their place among other events in a complex and chaotic network of inputs, outputs and feedback loops.

The trace for Derrida is *a way*, as in Heidegger, of destructing, reconstructing, an activity. It is a human activity, a way of connecting between that which can be experienced directly and that which cannot – bringing the transcendent into being as Being, as a process of mediation through a medium. Deleuze took a slightly different view when he introduced the plane of immanence – a virtual plane where all potentialities are real, ready to be actualized (as event) or not. For Derrida, things have to be made to happen – for Deleuze, they are already there as potential actualities.[12]

In order to fully understand the Deleuzian approach that Hainge develops requires the examination of a number of concepts central to Deleuze's thinking, particularly in relation to the *event* and the *virtual*.

The contemporary lived environment as mediated space invokes dimensionality and a non-linear understanding of events that are not simply superseded but remain always as echoes in the audiovisual archive – sometimes dormant, sometimes hyperactive, like musical notes ready to cohere with others in a multiplicity of ways to form scores; not written down but set free, never stationary but always fluid (Deleuze and Guattari 1987: 386). In *The Fold* (2006), Deleuze examines the event specifically in terms of an analysis of A.N. Whitehead that is apposite to the argument being formulated here. An examination of this analysis will serve to strengthen the move towards the acoustic and will theoretically underpin the case study to be developed in Chapter 5, as chaos is organized into harmonious events. 'Events are produced in a chaos, in a chaotic multiplicity, but only under the condition that a sort of screen intervenes' (Deleuze 2006: 86).

Deleuze points out that chaos for Leibniz can be approximated in the following ways: cosmological – the sum of all possibilities and physical –

depthless shadows, darkness, psychic – that equates to a kind of universal giddiness, as he terms it.[13]

All of these are countered, according to Deleuze, by a screen that creates something rather than nothing. It is a screen that filters the chaotic realm of limitless possibilities to allow compossibles, and only the best combinations of compossibles. It is a screen that illuminates the shadows, and as light contains all possible colours that can emerge as forms from the dark, it is a screen that orders perceptions from the infinite sum of all possible perceptions into those that we experience as events. As such, it is a screen that prevents us from, or saves us from, entering the quantum level of undecidability.

Chaos does not exist for Deleuze. It is an abstraction, an idea that comes to us already ordered via a screen that creates 'something rather than nothing'.[14] What does exist is a chaos that is constantly organizing itself. This perpetual state of organization, coherence and incoherence, of resonance and dissonance, includes individual beings who are themselves compossibles filtered through the screen.

But is the figure of the screen necessary? In a chapter by Celine Malaspina in Goddard *et al.*'s *Reverberations: The Philosophy, Aesthetics and Politics of Noise* (2012), the slightly different position of Ray Brassier is considered.

For Brassier, the totality of all compossibles is replaced with a totality of incompossibles – not a perfect universe in harmony but a noisy chaotic and potentially catastrophic one. In his *Genre Is Obsolete*, he argues that noise is the antithesis of genre and that it needs to be re-thought in order to recapture its subversive qualities. The ordering, filtering, stabilizing qualities of the screen are abandoned as the term noise itself becomes a magnificently unstable one. He says:

> 'Noise' has become the expedient moniker for a motley array of sonic practices – academic, artistic, counter-cultural – with little in common besides their perceived recalcitrance with respect to the conventions governing classical and popular musics. 'Noise' not only designates the no-man's-land between electro-acoustic investigation, free improvisation, avant-garde experiment, and sound art; more interestingly, it refers to anomalous zones of interference between genres: between post-punk and free jazz; between musique concrète and folk; between stochastic composition and art brut. Yet in being used to categorise all forms of sonic experimentation

> that ostensibly defy musico-logical classification – be they para-musical, anti-musical, or post-musical – 'noise' has become a generic label for anything deemed to subvert established genre. It is at once a specific sub-genre of musical vanguardism and a name for what refuses to be sub-sumed by genre. As a result, the functioning of the term 'noise' oscillates between that of a proper name and that of a concept; it equivocates between nominal anomaly and conceptual interference. Far from being stymied by such paradox, the more adventurous practitioners of this pseudo-genre have harnessed and transformed this indeterminacy into an enabling condition for work which effectively realises 'noise's' subversive pretensions by ruthlessly identifying and pulverising those generic tropes and gestures through which confrontation so quickly atrophies into convention.[15]

Here noise becomes a strategy or a musical, deterritorializing force, to return to Deleuze's description of the refrain from above. It mixes dissonant and often incongruous elements to enact a political strategy – not of opposition but of resistance. In Malaspina's chapter, Brassier's *Genre is Obsolete* is fused with Gilbert Simondon's paradigmatic method to produce a new form of engagement, an analogical form that produces a pattern where the ideas of coherence and incoherence that are both present in the term noise, can be used to explain other phenomena. In the case being developed here, the entangled refrains of order and chaos that are present in noise are similarly fused to make some sense of contemporary digital space.

Malaspina says:

> If we think of noise as a principle of differentiation (because it implies the radically new that is different from everything you are used to) and as index of heterogeneity (because no two forms of noise would satisfy this criteria of novelty if they were the same), then what kind of a ground is noise for an analogical relation? If the idea of a paradigm on the basis of noise is not a contradiction in terms, then what kind of a relation can it promise to illuminate between historians, performance artists, musicians, sound engineers, cultural theorists, etc.?
>
> (Malaspina 2012: 58)

This book, insofar as it proposes a relational ontology that positions scientific, political, aesthetic, economic, and cultural forces as relative to each other, hopefully goes some way to answering this question. Noise, as an eclectic analogical device for combining hitherto 'incompossible' realities, celebrates difference and excludes principles based on structural identity or resemblance.

But still the problem of Deleuze's screen has not been resolved. It is only from the order that it brings that we might discern the prospect of a lack of order – or a what if? But still we might ask, what is the screen? Despite its inclination towards the visual as a metaphor, the screen should be regarded as more of a sonic phenomenon. The screen for him is something like 'a formless elastic membrane, an electro-magnetic field or the receptacle of the Timaeus' (Deleuze 2006: 86). It is an electro-magnetic virtual screen that organizes chaos at the level of perception (a mediated environment), a plane of immanence without extension or actuality. This is the transcendental empirical realm – not yet revealed to experience but real nonetheless.

This visual analogy for Deleuze operates as an abstract machine that organizes chaos into order and is therefore *chaos media* in operation. For Brassier, however, it is too neat a solution. He prefers to endure and celebrate chaos and in his model there is no need for an abstract screen machine. Yet Deleuze is correct in identifying the fact that in actuality order does prevail.

To resolve this issue, if indeed we must, let us return to Hainge. He tries to fix what he calls the breaks in the thought of Michel Serres and Deleuze and Guattari, by positing noise – white noise – as the plane of immanence, an infinite ground – a machine that is not necessarily abstract and where the ontological and the phenomenological need not be separated. In so doing, he offers up an account of expressive becoming that has a relational sonic economy at its core and as such is useful in supporting the argument being developed in this book.

Hainge points out that, for Serres, noise had a strange ontological status: it is that which enables everything else to be, yet in itself remains only a potential in so far as when things come to be, noise itself ceases to be, making it a difficult concept to work with.[16] This is why Hainge re-fashions it into a workable idea via discussion of the abstract machine – a problematic concept in itself that he says can be corrected by fashioning out of noise a virtual plane that does not need an abstract machine to mediate between potential and actual. In doing so, Hainge develops Deleuze's assertion that the virtual is not opposed to the real but has elements of reality without extension that has not yet been actualized – it is an aspect *of* the real, something invisible that is transcendent and empirical. It is a generative agent of change.[17]

The concept of the virtual has also been explored by Frances Dyson in her book, *Sounding New Media* (2009), in which she states:

> The term 'virtual,' for instance, denotes the immateriality of digital media, while also referring to an ontological state or condition. As such, 'the virtual' contains within it the idea of a liminal domain wherein existence can take place and where users can 'be.' This domain is often described as transcendent, sublime, and mystical, uncorrupted by the social and political realities that dominate traditional media. The experience of this immaterial, simulated 'space' operates through 'immersion'—a process or condition whereby the viewer becomes totally enveloped within and transformed by the 'virtual environment.' Space acts as a pivotal element in this rhetorical architecture, since it provides a bridge between real and mythic spaces, such as the space of the screen, the space of the imagination, cosmic space, and literal, three-dimensional physical space. Space implies the possibility of immersion, habitation, and phenomenal plenitude.
>
> (Dyson 2009: 1)

The immaterial, the liminal and the transcendent in relation to digital space are concepts that are being challenged here. Digital space is not a separate place to be, not a place that needs a bridge connecting the real and virtual worlds, but a singular, relative, multidimensional, immanent space. As such, it should be contrasted with the speculative dualism as outlined above. For Meillassoux, the virtual abstraction is real – a realm independent of human perception of it, but real in a very different way from that which is being presented here. Ontology as relational is important here in terms of absolute and relative space and the character of digital space as always *in formation*, always forming as an expressive process, and doing so noisily (in formation as a verb – as opposed to having formed). If considered in this way, the use of noise as a means of understanding emergent order in a chaotic universe is distinct from the approach of object-orientated philosophers, such as Graham Harman who might identify in noise an 'undermining' of reality (Harman 2011: 7–19).

Returning to Hainge: 'It is then through noise, I will suggest, that we are able to intuit the serial relations that link the heterogenic modes through which everything comes to be in an ontology that does not believe in fixed identities, beings, and transcendent essence, but only difference, becomings and relations' (Hainge 2013: 14). Hainge approaches noise, unlike Prochnik, as a process and not as a result or as an effect. It does not bring about ends but keeps going. Hence in terms of its being oppositional in a political sense, one must rethink noise against the dialectical grain discussed above and rearticulate it in relation to the operation, he says, of an electrical field and the resistance therein. And this is where Hainge makes a very well-articulated point: 'The medium resists the transmission of the expression at the same time as the expression is entirely dependent on the system at the most fundamental level of base materiality, for its expressive potential can only be actualised in a material assemblage formed between the system and the expression that reconfigures both of them' (Hainge 2013: 17). This creates the impression of a complex environment, an electromagnetic political field where resistance and opposition are never dialectically resolved, or returned to rest, but are constantly thrashed out, giving rise to temporary truces and endlessly reconfigured agendas, in such a way that positivity and negativity become the medium through which each travels. It is an environment where without those who are opposed, the opposition makes no sense.[18]

To understand all of this is not an easy task because paradoxically it is in silence that noise matters, when as vibrating energy it moves, resonates, connects, coheres, forms – gives form to – before dissipating dispersing in chaotic dissonance. It is, as Deleuze might say, a virtual body without organs, an as yet un-realized entity that coexists with actualities and events that have already come to order. To think about noise in this way supports the idea of the musicality of the universe (virtual) where music is a harmonic ideal. 'Noise is the by-product of the event taking place' (Hainge 2013: 23). It is the energy produced by the virtual becoming actual, something that emerges on the phenomenological plane that is not necessarily unpleasant or subversive, sometimes music, sometimes not. Given that the virtual is always dynamic, always vibrating with energy, it is always noisy, especially when it is silent – producing waves both audible and inaudible, and always analogue and disrespectful of liminal thresholds between the real/virtual analogue/digital. This point is made by Steve Goodman in *Sonic Warfare* where he refers to Brian Massumi's assertion that: 'There is actually no such thing as digital sound, whether generated on or off a computer; if it's audible, it must be analog. Digital code is audible only after it is transduced into sound waves' (Goodman 2010: 117).

This was part of a broader attempt by Massumi to correct the erroneous conflation of the digital and the virtual (Massumi 2002). For Massumi, like Deleuze, the virtual is about potential whereas the digital is concerned only with predetermined, programmed possibilities. This is an interesting point in relation to the work of Leroi, as set out in the Introduction. It is also interesting in terms of the argument being put forward here that the analogue and the digital are not separate realms, but constitute a single relative space where vibrations travel comfortably backwards and forwards between the analogue and the digital, causing waves that are variously audible and inaudible as noise.

'Sonically speaking, then, there is no such thing as inanimate matter' (Hainge 2013: 1). Material bodies/objects coming into contact make further noise … But not just bodies. For Hainge, ideas and concepts also vibrate and resonate in relation to one another and with material bodies. As such, 'Everything, then, is expressive, not only embodying a form but for ever forming an embodiment' (Hainge 2013: 2). The idea of being expressive is central as a positive force/drive – the motor of change – a dynamism that creates the kind of noise that will be examined in Chapter 5. In short, everything is always moving.

This recognition of movement as an enduring quality of the physical is a central feature of non-representational thinking. In his *Non-Representational Theory: Space, Politics, Affect* (2008), Nigel Thrift stresses the recurring motif of movement and outlines what he regards to be the seven tenets of non-representational theory. A brief discussion of each, to demonstrate their relevance to the task at hand here will be useful.

First, he refers to *onflow*. This is the 'constant war on frozen states' and it follows a philosophical tradition that draws on William James and A.N. Whitehead. Thrift sets out his approach as a radical empiricism that goes beyond the sense-based empiricism of Hume but which stops short of removing the conscious subject from the equation altogether. As such, he is positioning himself, without acknowledging it directly, within the debate on speculative realism when he says:

> I want to temper what seem to me to be the more extreme manifestations of this lineage [radical empiricism], which can end up by positing a continuity of and to experience about which I am sceptical, by employing an ethological notion of the pre-individual field in which the event holds sway and which leads to 'buds or 'pulses' of thought formation/perception in which 'thought is never an object in its own hands'.
>
> (Thrift 2008: 6)

This quote may invoke the extent to which Actor Network Theory and, in a more radical fashion, speculative realism, push the pre-individual field to its limits to account for a realm of reality not accessible to human consciousness or perception. Thrift is less radical and clings, via James and Whitehead, to a kind of correlationism. As such, 'onflow' names a process of engagement where conscious individuals, despite forming part of a 'one substance cosmology', are never entirely irrelevant. Despite the conscious individual being significantly downgraded, Thrift is keen to avoid erasing them altogether. As such, he emphasizes play. Like Foucault, filtered through Philo above in Chapter 1, and Rancière, play is regarded as a significant site of activity, and it is one that figures prominently here, as the role of creative practitioners is examined from within a complex network of interrelated factors that aims to reconcile the relative significance of technology, politics, economics, and creativity or what Thrift refers to as a 'perpetual human activity'.

Onflow is a mobile concept that engenders movement as a central component. Thus, it requires that a temporal aspect be added to visual percepts, and a movement into acoustic space where everything that is, or can potentially be, encountered is subject to the as yet unknown. This relates to Meillassoux's account of 'Hume's problem' (2012: 82–111), that any predicted events cannot be known until they have been met: a problem that is too austere for Thrift. In terms of the argument being developed here, it is a problem that can be embraced rather than solved by invoking a sonic economy of unpredictable but nonetheless patterned and rhythmic harmonies that form, disperse and return as echoes in the 'curiosity shop' that is the lived digital environment.

The second tenet of non-representational theory is that it is *anti-biographical* and 'pre-individual'. Thrift questions the primacy of the conscious individual. As a tenet of non-representational theory, it again engages with issues raised by the speculative realists, particularly in relation to the role of the human. The question for Thrift, which is an important one, seems to be, how far can one go? If individuals are written out altogether, then certain political agendas can be 'smuggled in'. It is a question that is undoubtedly beyond the remit of this book, but one which needs to be highlighted, or amplified: if humans are themselves merely resonating events in a chaotic universe of chemical accidents, then horoscopes are probably as reliable as any physical or metaphysical method in guiding our planned trajectories. But maybe sound can become a more trustworthy indicator as some are beginning to hypothesize (Jacques Attali, and Jonathan Sterne, for example): 'Of particular interest to Attali was the ways in which modes of the sonic organization of noise in the form of music were not only explanatory of historically specific modes of power but even premonitory, as if noise was first organized sonically, before becoming organized socially and politically' (Goddard *et al.* 2012: 4).

If this is going too far, then perhaps sound, or a sonic economy, can operate as a kind of improvisatory plan emerges: not knowing exactly where one is going to go but making those decisions in time, in tune, in rhythm, or possibly missing the beat, the cue and getting lost in the chaos once more before finding Ariadne's thread and reconnecting with the refrain (Deleuze and Guattari 2004: 343).

Thrift goes on to say: 'Instead I want to substitute a material schematism in which the world is made up of all kinds of things brought in to relation with one another by many and various spaces through a continuous and largely involuntary process of encounter, and the violent training that such an encounter forces' (Thrift 2008: 8). This echoes the relational ontology of Hainge, and the backwards and forwards encounters between the digital and the analog as described above.

The third tenet is *practices*. Practices, like discourses, can be identified and analysed in order to make sense of the moment that we find ourselves in and to assess how they differ from previous moments as well as moments to come. Thrift describes this, saying:

> Material bodies are continually being rewritten as unusual circumstances arise, and new bodies are continually making an entrance but, if we are looking for something that approximates to a stable feature of a world that is continually in meltdown, that is continually bringing forth new hybrids, then I take the practice to be it. Practices are productive concatenations that have been constructed out of all manner of resources and which provide the basic intelligibility of the world: they are not therefore the properties of actors but of the practices themselves. Actions presuppose practices and not vice versa . . . as practices lose their place in a historical form of life, they may leave abandoned wreckage behind them (how does this relate to the idea of trace?) which can then take on new life, generating new hybrids or simply leavings which still have resonance.
>
> (Thrift 2008: 8–9)

Resonant 'leavings' or echoes work as acoustic refrains with disparate and uncertain forms coalescing to create discourses that, once identified, can melt into air, undergoing a quantum leap before reforming in a newer, or indeed an older space and time, or spacetime. To understand such complex patterning and re-patterning requires a methodology that does not rely solely on visual stimuli. According to such an approach (sonic economy), specific discursive phenomena that traverse what might be called mediated environments (where artefacts as things, words, concepts, percepts, or Bachelard's poetic images operate as media in what McLuhan has called an anti-environment that uncouples the figure–ground relationship to focus on the ground alone that is not present on the visual register (McLuhan and McLuhan 1988: 3)) can be described as having the characteristics of a statement without operating as exemplars or as fixed representations.[19]

Practices then become manifest at the point at which statements cohere and begin to reverberate and echo with the 'repeatable materiality' that Foucault identified, only now in a range of guises or forms that may not always be clearly visible. Hence they can be related to Lefebvre's idea of reappearances and their predictive function as previsionary portrait. (Lefebvre 2004: 19).[20]

The fourth tenet is *things*. Non-representational theory takes seriously the operational presence of things beyond their perception by humans. They stand distinct from human recognition (Harman 2002: 29, cited in Thrift 2008: 9), they operate in relation to other things, and they relate to the body *as* thing.

The fifth tenet is that non-representational theory is *experimental*.

> I want to pull the energy of the performing arts into the social sciences in order to make it easier to 'crawl to the edge of the cliff of the conceptual (Vendler 1995: 75). To see what will happen. To let the event sing to you. To some this will appear a retrograde step: hasn't the history of the social sciences been about attaining the kind of rigour that the performing arts supposedly lack?
>
> (Thrift 2008: 12)

This is what is happening here: events will sing to us. Social and political analysis will be inflected with creative force.

The sixth tenet is *affect/sensation*. 'All of this said, I do want to retain a certain minimal humanism. Whilst refusing to grant reflexive consciousness and its pretensions to invariance, the privilege of occupying centre stage, dropping the human subject entirely seems to me to be a step too far' (Thrift 2008: 13). Once more, this position is echoed here, but with the caveat that humans are themselves events that arise as figures from a ground, and not the sole arbiters of reality.

The final tenet is *space*. Where does all of this activity take place? This requires an engagement with the argument around absolute and relative space as was carried out in Chapter 1. It also requires, and indeed it initiated, the above discussion around the division between the real and the virtual, the ideal and the material.

What Thrift's succinct encapsulation of non-representational theory shows is that the question then of whether the future trajectory of humanity can best be understood in terms of discrete (digital), discontinuous and chaotic elements, or as a linear analogue unfolding towards an inevitable end point, is an enduringly important one.[21] Presently it requires that we reassess epistemological assumptions and in the process come up with new and relevant kinds of thinking capable of accounting for the perceived unpredictability of the digital age. This is happening in certain areas of philosophy and critical theory, yet does not seem to translate into any meaningful political action. In fact, existing economic, social and political models appear ill equipped to cope with the demands being placed upon them by advances in digital communication networks. This was evident in Chapter 2 where governments were shown to be actively renouncing any claim to determination. Instead these important issues are being addressed at the levels of creative practice and technological development.[22]

Combining cultural theory and technology is a project that has a long history. The Frankfurt School of critical theorists grappled with this combination tirelessly. Walter Benjamin said that technological advances in terms of the reproducibility of art changed its very function: 'Instead of being based on ritual it begins to be based on another practice – politics' (Benjamin 1936). This changed function is what this book is attempting to understand. It is trying to make sense of the place where we find ourselves: in a technological landscape struggling to resolve questions of the real and the virtual, the immanent and the transcendent, with creative practice – or art – as an activity that can help to understand and make sense of what is happening, in our space, our time.

Space, time and creativity are themes that were picked up by the pioneering artist and media theorist Nicolas Schoffer. He wrote:

> The real challenge in artistic research is to attain a maximum of freedom, freedom to define and select parameters for a combinative formula designed to enrich the quality and quantity of the effects produced. By developing ideas without redundancy and by allowing acute as well as unforeseen impressions to arise, one can thus explode the temporal and spatial limits of the work. In order to accomplish this in music, one must eliminate one of its essential characteristics: its very confinement within temporal and even spatial limits.
>
> Enclosing a musical work within a concert hall for a limited period with an equally limited public is as 'anti-socio-cultural' as enclosing visual works inside museums with limited spaces and entrance fees. But, while visual art, through architecture and sculpture, has broken down these barriers, music remains enclosed, if only inside appliances such as radios, record players, tape recorders or television sets. We must liberate music!
>
> (Schöffer 1983: http://www.olats.org/schoffer/eindex.htm [Accessed 10 Sept. 2014])[23]

This seems to speak to many of our contemporary questions. If the visual has escaped the 'anti-socio-cultural' spatial restriction in the form of architecture – insofar as it acquired temporality as a quality – then sonic structures have also begun to do so within technologically enabled environments.

This liberation of sound offers a means of accounting for a world that cannot be adequately represented in static visual terms and is a useful way of understanding the constant movement that characterizes the digital environment. It is a position that has been advocated by a number of theorists, not least Jean-Luc Nancy, and it is one that requires a reenergized kind of listening, one that is appropriate to contemporary conditions.[24]

Jean-Luc Nancy states:

> 'To be all ears' [*être à l'écoute*, to be listening] today forms an expression that belongs to a register of philanthropic oversensitivity, where condescension resounds alongside good intentions; thus it often has a pious ring to it. Hence, for example, the set phrase 'to be in tune with the young, with the neighbourhood, with the world,' and so on. But here I want to understand it in other registers, in completely different tonalities, and first of all in an ontological tonality. What does it mean for a being to be immersed entirely in listening, formed by listening, or in listening, listening with all his being?
> (Nancy 2007: 4)

This being 'immersed' speaks to our existence in relation to the information networks through which we constantly move. It also points towards a new kind of interactive and collaborative form of social and political engagement. Again it is art and technology that are leading the way, and again operate as a kind of play.[25]

In relation to this, Jacques Rancière pointed out:

> At the end of the fifteenth of his *Letters on the Aesthetic Education of Mankind* Schiller states a paradox and makes a promise. He declares that 'Man is only completely human when he plays', and assures us that this paradox is capable 'of bearing the whole edifice of the art of the beautiful and of the still more difficult art of living'. We could reformulate this thought as follows: there exists a specific sensory experience—the aesthetic—that holds the promise of both a new world of art and a new life for individuals and the community.
> (Rancière 2002: 1)

This emphasis on play, if we suppose it to be the case, places great demands on technology developers, academics and creative practitioners. It is a demand that goes beyond just economic gain and assigns them an enhanced social and artistic role. To understand these demands and to show how creativity can be foregrounded, requires that we challenge representational thinking. The realm of visual representation and its limited dimensionality then is a philosophical conundrum to which we must return again and again. We might do so by returning not only to Heidegger but also to Schopenhauer and his distinction between representation and the will and to Nietzsche and his Apollonian and Dionysian duality, where respectively sonority/music is a copy of the will and an intoxicating force that does not reveal itself on a visual plane. Music for both Schopenhauer and Nietzsche operates on an existential plane or in a parallel world where different rules apply and where alternative realities lie.[26]

But what if there are no alternative realities, but just one relational reality? If we are to fully understand the nature of our being in such a world and by necessity our current plight, we need to move out of what Marshall McLuhan called 180-degree visual space and into the 360-degree acoustic/sonic post-Euclidian realm. In relation to this, Frances Dyson has said:

> Sound is the immersive medium par excellence. Three-dimensional, interactive, and synesthetic, perceived in the here and now of an embodied space, sound returns to the listener the very same qualities that media mediates: that feeling of being here now, of experiencing oneself as engulfed, enveloped, absorbed, enmeshed, in short, immersed in an environment. Sound surrounds. Its phenomenal characteristics—the fact that it is invisible, intangible, ephemeral, and vibrational—coordinate with the physiology of the ears, to create a perceptual experience profoundly different from the dominant sense of sight. Whereas eyes have a visual range of 180 degrees, projecting from the front of the subject, ears cover a 360-degree expanse, hearing all around. Whereas eyes can be closed, shutting out unwanted sights, ears have no lids. Whereas seeing positions the subject symbolically as director of its look, always looking ahead toward the future, hearing subverts this role: the listener cannot control what is often overheard, what is muttered 'behind my back.'
>
> (Dyson 2009: 4)

Drawing on the acoustic/sonic in this way serves to support a sonic economy in terms of identifying links within a dispersed system of power that relies on its invisibility, its stealth, for its continued success. For Foucault, the formation of discourses and their constituent parts were not conveniently and quantifiably visible, they could not be simply mapped in gridded space that is absolute and static. Explaining how discourses operate in this way allows us to recognize them not only as regimes of power or as examples of dialectical progression, but as sonic components that configure and reconfigure in multidimensional spacetime. Their movement must be tracked across relative space in all its multidimensional forms.

In relation to this, Jean-Luc Nancy states:

> There is, at least potentially, more isomorphism between the visual and the conceptual, even if only by virtue of the fact that the morphe, the form implied in the idea of isomorphism is immediately thought or grasped on the visual plane. The sonorous, on the other hand, outweighs form. It does not dissolve it, but rather enlarges it; it gives it an amplitude, a density and a vibration or an undulation whose outline never does anything but approach. The visual persists until its disappearance; the sonorous appears and fades away into its permanence.
>
> (Nancy 2007: 2)

Foucauldian statements, like sound, traverse qualitative space – fading into permanence to one day be returned as echo. In so being, they can be identified along a diagonal vector somewhere between the vertical and horizontal axes of propositions and phrases and are mobile and atonal (Deleuze 1999: 3).[27]

Unconstrained by the isomorphic extended body, statements come into relation with one another in ways that are not restricted by spatial proximity nor constrained by notions of sensible similarity or difference. In this respect they are qualitative not quantitative.

On this matter, Leibniz said in his *Metaphysical Foundations of Mathematics*: 'Quantity or magnitude is that in things which can be known only through their simultaneous compresence – or by their simultaneous perception ... Quality, on the other hand, is what can be known in things when they are observed singly, without requiring compresence' (Casey 1998: 171). The absence of compresence is equivalent to Foucault's dispersion or Nancy's qualitative 'fading away into permanence'.

Statements move, echo, bounce, collide, connect and disconnect across space and time; not constrained by a knowing subject in Schopenhauer's representational field, not constrained by the body or the soundproof room but released beyond its extension in space – escaped from its solidity like the sound from the throat that the speaker does not recognize as their own (Lyotard 2002). Such formations are acoustic/sonic in ways both literal and non – literal, dealing specifically with sound but also by using sound as a non-representational model of organization for a multiple range of significant factors.

Such a model could arrange phenomena without fixing them in either space or time – allowing them to move in relation to one another, to create narratives without beginning, middle or end, to construct archaeologies of the present in the knowledge that at some future point they will need to be constructed anew. The patterns may not be neat, may not follow strict tonal logic, they may not even be patterns that are recognizable in any conventional sense, not ordered according to received notions of difference or similarity, not present in the corridors of power, not attributable to single individuals, not part of a consensual understanding. Yet they are there and they should be understood as being there without having to provide locational evidence. They are capable of evading such techniques and might only be knowable via recourse to, or a drawing upon, the sonic, acoustic, poetic realm of the artist who alone can create a necessary anti-environment. Yet the idea of the artist itself may need some considerable rethinking. This is why music of all kinds is important as an area of research that can function as more than just metaphor.

To pursue this task is to respond to Deleuze's comment that: 'If things aren't going too well in contemporary thought, it's because there's a return under the name of "modernism" to abstractions, back to the problem of origins, all that sort of thing . . . [modernism returned as an echo].' He continues: 'Any analysis in terms of movements, vectors is blocked. We're in a very weak phase, a period of reaction. Yet philosophy thought it had done with the problem of origins. It was no longer a question of starting or finishing. The question was rather, what happens in between?' (Deleuze 1995: 121).

What happens in between the three spaces, collateral, correlative, and associative, discussed in Chapter 1, is the mediation of statements, and it is a process that is in constant motion as they correspond with discourses. This involves engaging sometimes with invisible discursive formations or arrangements, statements that are neither compresent or tethered, that operate not in a Platonic transcendent or theoretical realm but in an immanent sensory one. Not only words and things and the causal relationship in either direction, but the inter-causal relationship back and forth. Visibilities, Deleuze tells us, must be *found* in *things* and the way in which we do this is crucial: it is the finding that is important here as a process. For finding implies something is hidden or not immediately present, like the identity of fairy tale characters as discussed above and cited by Sara Maitland. The manner of a coming into appearance, like the fire that casts the shadow in Plato's cave, may not present itself to the observer, thus language and form as medium and/or technology – or as technologies of power – contribute to the way things are lit but they themselves may not always be visible.

We need to find the ways in which statements move and form from the shadowy representations in Plato's cave to the digital realm. The way this is and has been done can be seen in archives that are themselves audio-visual as language lights up what it is we see and what we see gives rise to language formations in many new forms.[28] Contemporary developments in production storage and distribution of information bring statements into new relations with each other that are not amenable to static representation.

So the problem seems to be with representation and ontic, Apollonian, static states that cannot account for vibration, fading, and amplification (or reappearance as Lefebvre calls it). Beginning with the question of representation as a field of philosophical inquiry that impacts directly on the ways in which contemporary mediated environments are understood, sonority is presented here as non-representational and as embodying movement.[29] This idea is pivotal if we are to progress beyond what Heidegger calls the ontic, to the Ontological as a process that is always in motion and is capable of accounting for the fluid complexities of the contemporary age.[30] To do so is to engage with some fundamental philosophical questions that go back to classical Greek philosophy and its emphasis on visuality (McLuhan and McLuhan 1988: 4). It is also to emphasize the importance of motion in the analysis of sound and music that will be significant later in this book. Susan McClary, in an essay reflecting on Jacques Attali's *Noise: The Political Economy of Music*, argues that the tendency to reduce music to mathematical principles served in effect to render it paradoxically silent. She points to Jean-Philippe Rameau's tonal harmonic theory, based as it is on the primacy of the triad, as evidence of the tendency to reduce music to science and reason. She says:

> Breaking a piece of music down into a series of its smallest atomic units destroys whatever illusions of motion it might have had. It yields a chain of freeze frame stills. All of which turn out to be instances of triads. Mathematical certainty and the acoustical seal of approval are bought at the price of silence and death, for text, continuity, color, inflection, expression, and social function are no longer relevant issues. The piece is paralysed, laid out like a cadaver, dismembered, and cast aside.
>
> (McClary, cited in Attali 2006: 151)

This idea is relevant in terms of the continuous/discontinuous debate between Bergson and Bachelard, as well as to the digital and the analogue, to Leibniz and Locke, and more recently to Meillassoux's mathematization of matter in *After Finitude* (2012). And it presents something of a dilemma for the argument being developed here that proposes a discrete or digital conception of matter while simultaneously affirming the relevance of 'expression and social function' of sound and music as kind of continuous discontinuity. Sound and music must therefore be comprehended – or prehended as demonstrating physical qualities that occur naturally, and therefore mathematically, but in a way that recognizes movement and the refrain as an enduring quality (and quantity) – in short, to do so in a manner that is not bought 'at the price of silence and death'.

Notes

1 It is important here to consider the difference between Bergson and Bachelard in relation to continuous and discontinuous time – Bergson's image of time and his reflections on music, and Bachelard's discrete instants that can be slotted into place

to create new and original points of departure that are not in strict relation to a unity or totality [a continuity]. Bergson's reflections on movement and the role of the creative individual are also significant.

> Suppress the conscious and the living (and you can do this only through an artificial abstraction, for the material world once again implies perhaps the necessary presence of consciousness and life), you obtain in fact a universe whose successive states are in theory calculable in advance, like the image placed side by side along the cinematographic film, prior to its unrolling.
>
> (Bergson 1992: 93)

2 This process of reconstitution invokes the concept of data archaeology that is picked up in the following example: http://www.dataisnature.com/?p=1618 [Accessed 10 Sept. 2014].

3 Bergson presents the concept of duration as an indivisible unfolding that has movement at its centre. It is predicated on seamless transition and evolution. Any division in terms of time is artificial and arbitrary. Dividing lines between past, present and future are in essence not present – like individual notes in a song that merge each into the next. This perpetual flow defies mathematization, and it denies numerical value. In doing so, it releases the creative individual from the as it *is* as it *appears* dilemma as life as it is lived becomes an expressive and creative endeavour. For Bergson, the creative mind is not historically contingent within a static set of temporal conditions as perhaps it is for Foucault.

The seamless transitions lead Bachelard to the conclusion that if each point in the present is entwined with the past, then originality and creativity cannot reside there – the canon will always haunt the creative moment.

4 Refer to Harman's *The Quadruple Object* (2011).

5 This relates to Minkowski's well spring, in Bachelard (1994: xvi).

6 In contrast to Harman (2011).

7 Stuart Elden's Introduction in *Rhythmanalysis* (Lefebvre 2004: x).

8 Such a relational ontology is challenged by Harman (2010).

9 See Steve Goodman's *Sonic Warfare* (2010).

10 See Susan McClary's Afterword, in Attali's *Noise: A Political Economy of Music*, in relation to punk (2006: 149–58).

11 The intention here is not to oppose tonality but set it free (Deleuze and Guattari 2004; 386).

12 For a detailed exploration of the differences between Deleuze and Derrida, see *Between Deleuze and Derrida* (Patton and Protevi 2003).

13 'The event is a vibration with an infinity of harmonics or submultiples, such as an audible wave, a luminous wave . . .' (Deleuze 2006: 87).

See also:

> Deleuze's formulation of the 'individual' in *The Logic of Sense* is saturated with rigorous discussions on the relationship between the series, the resonance between them, the constitution of events, actualisation and counter-actualisation of the events, the disjunctive synthesis, and its affirmation, etc., which eventually lead to 'The Twenty-Fifth Series,' called 'Univocity.' The latter functions as a way of opening up the concept of 'individual' to all the compossible worlds so that the 'individual' will be organised within a multiplicity as an 'event.'
>
> (Zafer Aracagök: available at: http://www.rhizomes.net/issue19/aracagok.html [Accessed 10 Sept. 2014])

14 Perhaps a better solution is the one forged by Shaviro who appropriates Whitehead via Stengers as a means of correcting the Heideggerian focus on Dasein and essence. He urges us to ask, not why is there something rather than nothing but why is everything always new? (Shaviro 2012).

15 See http://www.toliveandshaveinla.com/genre_is_obsolete_brassier.pdf [Accessed 10 Sept. 2014].

16 'Noise is indeed static or interference but not that of an unorganized chaos so much as patterns of organization alien to the norms of a specific system – that which Serres refers to as "the parasite" ' (Goddard, *et al.* 2012: 3).

17 For a detailed discussion of the possible and the real, see Shaviro (2012: 16–46).

18 This is a similar argument to the one proposed by Joseph Nechvatal (2011) via his use of the Deleuzian Vacuole as an element that is central but not distinct from a system of production or expression (Nechvatal sees art operating in the same way – a kind of anti-environment or abstract machine).

19 Foucault's refusal to be drawn into providing examples was an important stance that highlighted the fluid nature of his method that must always be in process and not static in terms of citing fixed examples against which all other cases might be judged. Such a generic taxonomy of statements would be so busy allotting phenomena to their rightful place that it would miss the formation of new discourses as statements moved out of view or as they morphed within and between strict categories.

20 According to Lefebvre:

> In general, one does a portrait of someone who exists and who tempts the painter, the novelist or the playwright. Is it possible to do a portrait of someone who does not yet exist and which would have to help to bring about his existence? Yes, if one finds the traits inscribe themselves on a face of the future, which will cast aside false resemblances, thus enabling us to foresee the dissimilarities.
>
> (Lefebvre 2004: 10)

This can be related to the coherence of statements that were identified in Chapter 2, as evidence of dissimilarities coming together to map out the future shape of communication networks.

21 See Floridi (2009), available at: http://philsci-archive.pitt.edu/4076/1/ado.pdf [Accessed 10 Sept. 2014].

22 What kind of political function can be assigned to art and creative practice in a contemporary setting? A good example is the work of the artist Rafael Lozano-Hemmer, see http://www.lozano-hemmer.com [Accessed 10 Sept. 2014].

23 For an historical account of sound art and its relationship to visual art, see Kim-Cohen (2009).

24 Such as Pauline Oliveros, Sara Maitland, and Les Back.

25 Dyson argues that the use of aural metaphors has been applied to new media without a full understanding of the phenomenology of sound. This is problematic insofar as:

> By using terms such as 'immersion,' it is possible to conceal all the necessary social and technological interventions and delimitations that go into producing digital media, by foregrounding the all-embracing, all-enveloping condition the term suggests. The use of 'immersion' in the title thus signifies both immersive technologies and the desire to elide the distance that technological mediation imposes—between the user and the apparatus, the real and the simulated, the natural and the artificial, the human and the technological.
>
> (Dyson 2009: 7)

She argues that it has been used to cover its own tracks: 'Fundamental to the development of communications technology, these notions are revised, remodeled, cut, and pasted to fit the new techno-epistemological regime, and, as I argue throughout the book, they are never far from the influence of sound—either as a medium, or a model, or a metaphoric ground' (Dyson 2009: 7).

26 See Schopenhauer (1966) and Nietzsche (1967).

27 For a discussion of propositions and phrases, see Foucault (2004a: 119–32).

28 The description of the archive as audio-visual is important here in justifying the wide range of sources/statements that one might draw upon in order to identify a discourse.

29 Movement is a leitmotif of non-representational theory (Thrift 2008: 5).

30 The digital information age is itself a discourse formed of statements born of Western scientific rationalism with echoes of classical liberalism and the hippy ideals of the 1960s. It is emitted and returned as it spreads. As we *watch* the second world industrialize in our own 'image', however, we run the risk of missing something. The sound of science has emanated from the West and has resounded in the East before rebounding as an echo of vindication, reflected; self-congratulatory and isomorphic.

4

An Invisible Exchange

Before establishing the framework for a sonic economy, it is important to clear up once more some conceptual terms. In this case, Deleuze's *difference* and *repetition*. To do so will allow the idea of an invisible exchange to be elaborated. For Deleuze, *repetition* stands as distinct from the general that is prefaced upon resemblance and equivalence (exchange). If something can be repeated, it has no equivalent, it is a singularity and cannot therefore be exchanged. As in the case of festivals that repeat:

> To repeat is to behave in a certain manner, but in relation to something unique or singular which has no equal or equivalent. And perhaps this repetition at the level of external conduct echoes, for its own part, a more secret vibration which animates it, a more profound, internal repetition within the singular. This is the apparent paradox of festivals: they repeat an 'unrepeatable'. They do not add a second and a third time to the first, but carry the first time to the nth power.
>
> (Deleuze 2008: 2)

What is exchanged then are generalities, and as will be demonstrated later, these have qualities that are invisible.

Difference, on the other hand, should be seen as a positive force – a force of distinction, not a measure of the *difference between*. The significance here is in relation to economic processes of exchange – generality of the head and repetition of the heart – and how economies are understood and how patterns are formed.

The significance is also in relation to comparisons between Coventry and Detroit that will be carried out in Chapter 5. At the level of resemblance, certain conclusions can be drawn, but when new kinds of difference and repetition are engaged other conclusions emerge. This book is less interested in the differences between or in the resemblances, and more interested in the intense singularities that repeat like festivals, the repetition that cannot be exchanged but which can occur in various locations in space and time, in this case the repetition is the force present in music that is able to resist without opposing. It is a repetition that opposes racial division and the dismantling of industrial communities.

The mediated environment of digital space as defined above is a shadowy mysterious place. Proceeding from the notion of mystery and demystification in *Money Bags Must Be So Lucky*, Robert Paul Wolff sets out a unique position in relation to political economy. He begins with the often stated position, recognized in the thinking and writing of Marx and Engels, that the Germans were essentially philosophical (religious), the French were essentially political and the English practical (economics). The first, Wolff says, had been demystified by Kant and Feuerbach, and the second by the French Revolution in the eyes of Marx. However, when he approached the practical economic disposition of the English, he encountered a phenomenon that common sense dictated did not need demystifying. This was a realm already transparent and rational: 'The political economists who recorded and anatomized the doings of the marketplace reflected this simplicity, this absence of pendulous transcendent significance, of shadow and echo' (Wolff 1988: 41)

Adam Smith, David Ricardo, James Mill, Jeremy Bentham, John Stuart Mill: all of them wrote clearly without recourse to metaphysical posturing about a situation that was clearly defined and understood as regulated equivalences of exchange and contracts enforced by law. However, when Marx looked more closely, he found that the terrain of markets was deeply complex and mysterious. Indeed, it was more so than the philosophical and political arenas that had previously been addressed: 'The principal verb of the opening sentence of *Capital* is *erscheinen als*, "appears as." Marx chooses to begin his analysis of capitalism, and of the theories of capitalism advanced by his predecessors, at the level of appearances, thereby invoking the distinction between appearance and reality on which his entire theoretical enterprise depends' (Wolff 1988: 44).

It is the same distinction on which the theoretical enterprise embarked on in this volume depends to a large extent: things as they are and things as they appear. The question of materiality was key for Marx and Engels. Engels famously wrote of this matter in relation to what he would come to describe as the two faces of Manchester. In a manner that should lead us to reflect on the work already done in this book in relation to space and place, Engels describes a Manchester as it might be experienced by the 'money aristocracy', a Manchester that circumvents for anyone of that class the dirty reality of manual labour – hidden by way of façades and routes intended to take merchants to their destination without having to see the reality, and a different Manchester that was abuzz with the daily grind of labour. Which was real? That which appeared or that which was hidden? In turn, Marx took up the question of reality. This time in relation to the nature of a commodity: 'Indeed a commodity strictly speaking is not an object of sense perception at all' (Wolff 1988: 46). So what is it?

A commodity is a category of things that possess characteristics beyond their use value and beyond the value of the labour expended in their production. They acquire metaphysical qualities that are not necessarily rational, and not necessarily visible but real – *crystals of social substance*. As such, they might be regarded as being configured in terms of a noisy materiality. In a way, what is happening here is that out of actuality comes the virtual (Deleuze inverted this as Marx himself had inverted Hegel). Once a thing becomes a commodity, it is no longer sensible (in a way it makes no sense), its thingness dissipates into abstraction. A thing can be perceived but a commodity cannot because it has qualities that cannot be sensed (another inversion of Hegel). It starts life as a material thing and becomes immaterial as commodity.

Hence any notion that the market as a realm of exchange is real is questionable and this has implications for those commentators who argue in terms of a move from real to virtual economies. Marx had already rendered such a move unnecessary: 'The quantum of exchange value congealed or crystalized in each commodity can neither be seen nor felt nor smelt, nor tasted' (Wolff 1988: 47). But can it be heard?

Marx inverts not only Hegelian thinking but also that most classical of idealist tropes: Plato's Allegory of the Cave. For Marx, the brightly lit marketplace is an appearance that disguises the dimly lit reality that resides behind the factory gates (the cave): 'It is in the production of commodities, not in their exchange, that the secret of capitalism lies' (Wolff 1988: 53) The exchange is invisible.

How does this relate to digital space? For some, the digital marketplace is essentially different to its real predecessor: 'Its drives and the extent of its fluctuations stem from its unreality, its de-realization. This is because the new economy is based on the production and exchange of information and no longer so much on merchandise – bits not atoms, says Negroponte' (Fischer 2006: 152). From this perspective, it is regarded as an imaginary space:

> The dematerialization of the economy encourages not only its fluidity and acceleration, but also its volatility and, consequently, the grip that the imagination can have on it. The value of a loaf of bread or a house is easily ascertained; the object is there visible and perceptible. But what is a concept worth? What is a fantasy, or a vision worth? And where is it?
>
> (Fischer 2006: 153)

Again here is a privileging of the visual and the spatially extended as a real and knowable object. But as Hainge pointed out and as discussed above – why should bodies (extended) have all the fun? Concepts and fantasies also vibrate just like bodies (nothing is inanimate).

Fischer says that the imaginary real that is the e-economy (or i-economy) has a religion. It is Ultraliberalism – an economic law of nature that eschews any form of intervention as consumers replace citizens in a global free-for-all, and where cocooned individuals have their isolated sadness alleviated by the presence of manufactured utopias in the shape of hybrid shopping centres and entertainment complexes.

> Marx said that we would finally achieve happiness after the communist revolution. He embellished the myth by suggesting that there would be great heaps of merchandise from which each person could draw according to his or her needs. Now Marx's promise appears to have been fulfilled by e-commerce and the major shopping centres of the new economy, which have brought together the two driving forces of American democracy: the consumer industry and the entertainment industry, the religion of consumption and the cult of distraction.
>
> (Fischer 2006: 157)

However, such market-driven democracy would be as difficult and unlikely to achieve as Marx's social justice, according to Fischer, who, despite championing the new virtual economy, does insert an element of caution. He challenges the polarized dualism of the real and the virtual, advising a reconciliation between the two as they shift on their axes in relation to one another. The difficulty arises because, as Henri Lefebvre pointed out, the visible moving parts are in danger of obscuring our view of the machinery. What is immediately visible may not tell the whole story

An adequate assessment of the impact of information technologies in terms of obscuring our view of the machinery cannot be derived simply from an analysis of the debate on technological advance in a modern society, however, but must work through a relationship between technology and economics, politics, and culture. Such posturing on technological advance, as it became evident during the last two decades of the twentieth century and into the twenty-first, and that heralded the move into digital space, concentrated on the concepts of information and the medium of exchange, which in turn are at the very core of any notion of political economy.

Developments in the media realm have long been used to support claims related to the shifting nature of society (Vattimo 1989; Kellner, cited in Raboy and Bruck, 1989: 131–46). Collins and Murroni, in their book, *New Media, New Policies*, have pointed out: 'Media and communications have never been thought more important. Pervasively they are seen to be key agents of change' (Collins and Murroni 1996: 1). The timing of this comment reflects the crescendo of enthusiasm for 'new technology' that was audible in the mid-1990s, as discussed in Chapter 2.

The theoretical position adopted when assessing the climate of claims and counter-claims in relation to technology are key in terms of perspective. At various moments, claims relating to the new technologies have been influenced by liberal free market thinking, or Fischer's Ultraliberalism, which abstracts technology from the economic and political processes and views it as essentially neutral, or by critical theory which moves through a Marxist preoccupation with the material to embrace notions of discourse (Poster 1995: 57). But what do we mean when we use the term 'technology'? The basic proposal is that while technological developments in a material sense can easily be identified, a thorough understanding of it and humankind's relationship to it cannot be satisfactorily achieved without considering the notion of technology in its other sense. That is the sense in which Foucault considered technology – in terms of specific relations of power and as inseparable from wider social discourse or practice (Poster 1995: 67). In this sense, the term is used to connote methodologies, including language, employed in a 'polymorphous' struggle for domination and control that might be said to constitute a challenge to the dialectical process in terms of the debate surrounding technology (Heim, cited in Lunenfeld 1999: 25).

Embracing this approach means that specific discourses can be linked to, and seen as intertwined with, technology that in turn can be viewed as a concept that transcends the limits of its material manifestation. In light of this, it is important to consider whether issues beyond the material manifestation of technologies serve to inform our understanding of them. Continuing to address the question, 'what is technology?', it is also useful to consider the two definitions as they were set out by Martin Heidegger: that technology is either described as a means to an end, or alternatively as a human activity. Heidegger collapsed this distinction in a manner that is relevant to this book and to the arguments that seek to collapse the cause-and-effect distinction between the media technologies and the political process. He suggested: 'The two definitions of technology belong together. For to posit ends and procure and utilize the means to them is a human activity' (Heidegger 1977: 4).

To assist us in our task, it is worthwhile examining the assertion that:

> Technology is not equivalent to the essence of technology. When we are seeking the essence of 'tree', we have to become aware that that which pervades every tree, is not itself a tree that can be encountered among all other trees.
>
> Likewise, the essence of technology is by no means anything technological. Thus we shall never experience our relationship to the essence of technology so long as we merely conceive and push forward the technological, put up with it, or evade it. Everywhere we remain unfree and chained to technology, whether we passionately affirm or deny it. But we are delivered over to it in the worst possible way when we regard it as something neutral; for this conception of it, to which today we particularly like to do homage, makes us utterly blind to the essence of technology.
>
> (Heidegger 1977: 4)

Heidegger was interested in the relationship between technology and the political process and particularly the way in which a political programme might be enacted to accommodate technological advance. He said: 'It is for me today a decisive question, how in general a political system – and which one – can be adapted to the technological age' (Pangle 1992: 47).

The search for a political system, or a political economy that can accommodate new technological developments when allied to the search for the essence of technology which includes consideration of associated political discourse, leads us to reject assumptions based on a liberal cause-and-effect model and to seek further critical engagement with the relationship between politics and technology in an attempt to fully understand the democratic implications (Pangle 1993: 34–47).

What this brief engagement with Heidegger serves to do is to stress from a theoretical point of view the inadequacies inherent in any approach which aggregates the 'technological' from wider social processes. In terms of perspective, despite having a tendency to see technology as a totalizing force, and possibly overstating its potential effect, Heidegger encouraged a view of technology that went beyond the material surface and included consideration of the related socio-political process being examined here. The intention is to demonstrate the complex relationship between material development and its utilization along social political and cultural lines (Edge, cited in Heap *et al.* 1995: 14–28). Or put more simply, to challenge notions of technological determinism and claims made in its name relating to the demise of the nation-state and the political process more widely.

It is therefore essential that one looks not just at the material manifestation of technology, but that it is assessed as part of a wider process, an economy, which includes the material objects themselves, the way in which they are utilized and the decision-making processes that have contributed to that utilization, as well as the discourse process which serves to normalize practice. It is an approach that is compatible with Actor Network Theory (Latour 2005), and with certain, but by no means all, aspects of speculative realist and object-orientated approaches. But it is also similar to an earlier manifestation of systematic thinking in relation to technology specifically, as proposed by J.D. Slack in her book, *Communication Technologies and Society* (1984), that emerged at a time when technology, as was seen in Chapter 2, was beginning to occupy intellectual thought once more. But the approach being developed here is different in one particular respect: whereas Slack placed technology as object within a structural framework, with the economy as the dominant force in accordance with her totalizing Althusserian model, here the object is set against relative post-structural thinking and considered in tandem with the concept of 'technologies of power' (Slack 1984: 89–91). This allows the specific material manifestation of technology to be viewed as having been created and developed within a political economy characterized by a complex set of competing discourses. Assessing technology in this way allows any totalizing tendency to be overcome.

Going back even further, Raymond Williams also engaged with the relationship between political economy and technology when he wrote:

> [A]ll questions about cause and effect, as between a technology and a society, are intensely political. Until we have begun to answer them, we really do not know, in any particular case, whether for example, we are talking about a technology, about a necessary institution or particular and changeable institutions, about a content or about a form.
>
> (Williams 1990: 10)

This notion of cause and effect as between technology and the political and economic process is being examined specifically here. Williams warned against the dangers of technological determinism, saying: 'It is an immensely powerful and now largely orthodox view of the nature of social change. New technologies are discovered, by an essentially internal process of research and development, which then sets the conditions for social change and progress. Progress, in particular, is the history of these inventions, which "created the modern world"' (Williams 1990: 11).

He further identified two dominant approaches. First, technological determinism, wherein technology is viewed as an internalized process that operates according to its own logic, and which then has a particular effect on society. Elements of this approach will be identified below when the liberal/free market position is examined.

Second, he discussed what he termed the symptomatic approach. In this case:

> [Technology] becomes available as an element or a medium in a process of change that is in any case occurring or about to occur. By contrast with pure technological determinism, this view emphasizes other causal factors in social change . . . It only acquires effective status when it is used for purposes which are already contained in this known social process.
>
> (Williams 1990: 13)

These approaches were also examined by Slack who categorized them as *simple* and *symptomatic* approaches (Slack 1984: 53–63). Both abstract technology from socio-political forces. As either a symptom or a catalyst; as either cause or effect. In either case not *deliberately* created and shaped.

This raises the question of the role of the subject and intentionality that often serves as the basis for criticism of Actor Network Theory.[1]

Raymond Williams suggested a renewed approach that sought to view technology as part of wider socio-political processes (1990: 14). It was a position that was echoed by David Lyon who suggested that technology 'can be shown to be the outcome of social shaping itself, including certain deliberate political, economic and cultural choices' (Lyon 1988: 157). Hence any analysis of technology and its related effects must look beyond the technology itself (Finnegan, Salaman, and Thompson 1987: x). It must look beyond the object itself. Slack described such an approach in terms of 'Expressive Causality' (Slack 1984: 64–80), viewing technology as an expression of the wider social process, as part of a totality that demonstrated essential qualities. Therefore, any change in the nature of technology could only occur as a result of a change in the essence of the totality. Such an approach, while interesting, is rejected here as technology is granted relative autonomy within an established system that is seen as having influenced technology without being the *sole* determinant.[2]

Numerous claims were made regarding the social and political effects of new information technology as the digital turn reached its apogee in the mid-1990s. The following will critically assess some of these claims as they occurred at the level of intellectual debate – representing a specifically located discourse. Employing two broad categories – the liberal/free market, and critical theory – the claims made by the various commentators are evaluated in relation to their theoretical position. For, as has already been mentioned, the position of the viewer can fundamentally affect the interpretation of events. Depending on which view is taken, any newly developed technology can be viewed as a potentially liberating force within a liberal free market environment, as being determined by the ruling economic class, or as being the subject of a constantly shifting power struggle and as having negated any lingering faith in dialectical materialism.

However, concentrating on perspective may raise a problem. Any attempt to reveal the truth in relation to technology is compromised by observation. As with Schrödinger's cat, it may exist in potentially different states, and only be fixed as a phenomenon when observed. Or as Harman might say, when the object has been overmined (Harman 2011: 7–19).

But having noted this word of caution, let us continue. First, with the liberal/free market approach: such an approach aggregated technological developments within the new media environment from wider social/political forces. It suggested generally that the free development of technologies had come to fundamentally affect the reality of social organization. Exponents of this approach suggested that populations would live in an interactive information-based techno-sphere of virtual debate and democracy, or possibly even beyond democracy as this was then perceived as being practical (Martin Walker, in *Wired*, April 1995: 43). Some observers were optimistic and others were pessimistic, but what they had in common was the certainty that technology had delivered humankind unto a particular point in history, or even beyond history altogether (Fukuyama 1989). Much of this certainty and inevitability was based on the work of Marshall McLuhan. He suggested that: 'The alphabet and print technology fostered and encouraged a fragmenting process, a process of specialism and detachment. Electronic technology fosters and encourages unification and involvement' (McLuhan 1967: 1). Echoes of this perspective are evident in many of the contemporary discourses on technology. The vision of global interactive unity is one that is said by some to be finally coming to fruition. The presumption has been made that technological developments would have a great and sudden predictable effect. Like McLuhan, such commentators assign technology essential and inevitable determining qualities. However, if the theoretical framework for viewing technology that has been presented here is adopted, then the role of the political process can be seen as central in shaping that technology, which in turn might be seen to affect societal change.

Perhaps the most visible example of this liberal/free market position could be seen in the pages of *Wired* magazine – occupying, as it did, a space between academic and popular discourse – with its McLuhanesque view of technology aligned to a strong belief in the principles of classical liberalism. It quoted directly from McLuhan. Over five pages adorned with technographics set out to visually represent its vision, using both English text and Chinese characters, as an indicator of global trends (http://www.rewired/96/fall/1118.html [Accessed 8 June 2014]), it said:

> The medium, or process, of our time – electric technology – is reshaping and restructuring patterns of social interdependence and every aspect of our personal life. It is forcing us to reconsider and reevaluate practically every thought, every action, and every institution formerly taken for granted. Everything is changing . . . Your family, your education, your neighbourhood, your job, your government, your relationship to 'the others'. And they're changing dramatically.
>
> (McLuhan 1967, from *Wired*, Premiere UK edition, 1995)

This was seen as having particular contemporary relevance in 1995 and thus formed part of a discourse as it had begun to proliferate around this time, as previously academic concerns began to enter the popular realm. The magazine stressed the somewhat contradictory view that, having been freed by technology, humankind now had the opportunity to assert its freedom. McLuhan's ideas were being reborn. The magazine announced from its cover, which featured a portrait of Thomas Paine: '*We have it in our power to begin the world over again*.' And, next to a smaller portrait of Paine, proclaimed him the patron saint of the masters of the infosphere. It said: 'We see with other eyes; we hear with other ears; and think with other thoughts than those we formerly used' (*Wired*, Premiere UK edition, 1995). This view credited technology with the power to enable a liberal utopia. The technology in question was viewed as fundamentally neutral: free from political/ideological influence. Essentially, technology was being viewed in a manner that ignored Raymond Williams's earlier emphasis on the relationship between technological development and wider socio-political forces. It was thought to have specific characteristics, which, if allowed to develop unchecked, would incline it towards certain ends. The only options in terms of human agency were those of how to respond. Allow it to take its natural course embracing all that it had to offer, or try and control it thus limiting its potential. Either way, the technology itself was seen as the determining factor. However, technology may not possess specific traits, but rather may have them assigned by a dominant political will (Lyon 1988: 157).

Not all commentators were as optimistic as those writing in the pages of *Wired*, however. Ian Angell in his article entitled, 'Winners and Losers in the Information Age' (1995), identified negative technological trends that he interpreted as causal factors in a process of fundamental global change, as discussed above.

This approach demonstrated equally blatant characteristics of technological determinism and the idea that the arrival of new technologies could and would facilitate fundamental change. The idea that these new developments may be the result of, or subject to, the demands of existing power relations was never placed very high on the agenda. Whether utopian or fatalistic, the anarcho-capitalism of their vision was shared and linked by a classical liberal belief in competition and the freedom of the market. They saw the world as being liberated from the shackles of state control. These ideas have been challenged here in a manner that places political discourse at the centre of the digital turn.

If a liberal approach views technology in the contemporary sense as a generic universal principle, it never went as far as uncovering its essence. This was due to the aggregation of 'technology' from wider forces, institutions and practices. An alternative approach was proposed by commentators who adopted a more critical position – from Marxism to post-structuralism. Beginning with a Marxist stance, it has been claimed that: 'Scientific advances would serve not the integration of human species at a higher, more automated level, but would be limited by constraints of the process of capital accumulation' (Poster 1984: 62). This view suggested that technology developed within the framework of class relations. It is an important point in considering whether technology in itself can have tendencies contrary to those assigned to it by the dominant social group. According to this model, technology would develop for the benefit of the ruling elite until such a time as the contradictions in the relations of production become so acute as to prompt revolutionary change. The question here, however, is, does information technology in its contemporary manifestation fit into this model? Is it the product of material labour – a human activity, or a contributory factor in negating the primacy of labour and the subject–object relationship that is central to Marxist thought? In answering this question, Alex Callinicos identified the phenomenon of the commodification of information. However, unlike Lyotard and the postmodernists, Callinicos saw it not as having negated the grand narrative but as having been central to it. If information was a commodity, then control of it represented little more than an expansion of the ownership of the means of production. When viewed from this point of view, one's attention shifts from the technology itself and its inevitable effects as discussed above in relation to the liberal approach, to an analysis of the wider relationships of power connected to the development and use of technology.

In challenging the liberal tendency to abstract technology from social and political factors, and to counter their claims relating to a technologically driven movement into a fundamentally distinct period of history, most often referred to as postmodernism, but also described as post-industrialism or the information age, it is useful to consider further the ideas of Callinicos, who said: 'I do not believe that we live in "New Times", in a "post industrial and postmodern age" fundamentally different from the capitalist mode of production globally dominant for the past two centuries. I deny the main theses of poststructuralism, which seems to me in substance false' (Callinicos 1991: 5).

For postmodern thinkers, particularly those who trumpeted the triumph of liberalism and the coming of a new age, the central element of their thought was the notion that significant change had occurred in economic, social and cultural terms; to the extent that many older ideas were no longer relevant. Callinicos took this to mean that primarily socialist ideas were seen as no longer being relevant. In order to support their refutation, many thinkers turned to and relied on the idea of the post-industrial society, one characterized by a significant decline in manufacturing brought about by developments in the area of information technology (Poster 1990: 21–42) – an idea popularized by people such as Daniel Bell (1974) and Alain Touraine (1971), and echoed later in the work of those such as Manuel Castells (1989), Heller and Feher (1988) and Ian Angell (1995), as discussed above. This motif of industrial decline forms the basis of the argument that will be developed in Chapter 5 in relation to the decline of the one-time industrial giants, Coventry and Detroit.

Callinicos was also critical of those who observed a fundamental cultural shift within post-industrial society; particularly Jean Baudrillard, of whom he wrote: 'The cultural phenomena on which others concentrate are for Baudrillard mere symptoms of a more fundamental change, which deprives us of the ability to talk of a world independent of our representations of it, to distinguish between true and false, real and imaginary' (Callinicos 1991: 145).

Callinicos rejected all of this, refusing to view the cultural realm as abstracted from wider social and political forces, and as having the ability to act independently, as Daniel Bell (1974) and Castells (1989) had suggested.

Such a Marxist approach viewed humankind's mastery over the physical world as fundamental to all history and therefore social development. All technological innovations would necessarily be subject to utilization as determined by the relations of production. This was the view of Callinicos and is one that employed the totalizing concept of historical materialism, collapsing specific developments in information technology and the notion of the 'information society' into a classical Marxist narrative. As such, it was evocative of Raymond Williams's 'symptomatic approach' in so far as it regarded any new developments as being the result of capitalism's inexorable progress. At the centre of this model lay the belief that information workers were involved inextricably in the productive process and on-going class divisions (Inglis 1990: 21). Again technology was seen as a generic concept, this time subject to economic and material interaction. But unlike some of the ideas above, it attributed technology with no power to affect change over and above the existing relations of production. Hence any notion that those developments signalled the progression into a new historical period or epoch was rigorously denied.

This notion of historical materialism and the idea that labour represented, in terms of the relations of production, the blueprint for social organization was questioned by Michel Foucault. He challenged Marx's insistence that labour represented humankind's primary activity. For him: 'The first assumption in Marx's texts that needs to be questioned is the notion of human beings acting upon nature. Marx constitutes the social field as one in which human beings act upon natural materials to produce natural objects' (Poster 1984: 46). There was a shift here from specific economic relations of power, as supported by Callinicos, to a far more heterogeneous approach that saw power as dispersed beyond the material and economic, across a far wider social field and as being encapsulated in and disseminated by means of the category discourse/practice, a category which suggests that non-discursive elements (practice) cannot be separated from the concept of discourse itself (Poster 1995: 82). The interplay of language was seen as central, over and above material relations of production. According to this point of view, in a manner similar to that discussed in relation to Heidegger above, material elements – in this case technology as object or thing – could not be adequately assessed without recourse to the discursive regimes to which they were connected. Foucault suggested: '[T]here are manifold relations of power which permeate, characterize and constitute the social body, and these relations of power cannot themselves be established, consolidated nor implemented without the production, accumulation, circulation and functioning of a discourse' (Foucault, cited in Gordon 1980: 93). Such discourses for Foucault were dominated by the state, enshrined within law and were best identified in written texts which could be seen as transcending the individual subject and as being intertwined with practice. 'The value of Foucauldian analysis rests with the conviction that the close reading of scientific discourse may uncover language patterns which, when associated with practices, position those practices in definite ways and legitimize the patterns of domination inherent in those practices' (Poster 1990: 89).

He thus aided our progression beyond the Weberian stress on action and the Kantian notion of scientific knowledge as neutral and independent in relation to the political process and ideology (Poster 1990: 88). In relation to this, the current proliferation of digital technologies can be described as having developed within an environment characterized by a dominant discourse which displayed internal complexities and which had discernible external effects, as demonstrated above.

Foucault pushed the analysis of technology beyond its material manifestation, placing it within a matrix of power. He saw power as localized and specific: not as part of some totalizing structure but as part of a relative multitudinous network of complex social relationships. Hence, we need not view information technology as an integral part of a totality. Indeed, Foucault, in that he was responsible for 'breaking with the totalising ambition of the universal intellectual' (Kirtzman 1988: xiv), assists us in a specific examination of one particular area of technology. From this point of view, the development of communications technology can be viewed as a specific phenomenon, and the claim can be questioned that it contains within it the impetus for universal or epochal change. There is in this area evidence of continuity, as opposed to the ruptures or discontinuities that Foucault identified as central to epistemic shifts, in terms of bureaucratic structures as well as in official thinking. This further highlights the usefulness of a Foucauldian approach, allowing us to assess the validity of claims relating to fundamental change. What became evident through the analysis of specific documents in Chapter 2 was the presence of a continuity of approach at the discourse level at least, particularly in relation to information networks and their position within a new media environment.

None of the above approaches alone are satisfactory. There is little evidence of libertarian trends, Marxist assertions about the dominance of a ruling class seem outdated, and the Foucauldian discourse model may be too dismissive of political economy and the attempt to reinvigorate historical materialism, as it was carried out by Habermas. But using a Foucauldian approach does allow us to assess political posturing as an element central to any notion of technology and to assess the extent to which it exerted power in the form of discourse (Poster 1995: 87–8). If we are required to change our perspective with regards to technology itself, the temptation to attribute to it the ability to affect fundamental social change should be guarded against. There has been change, but not to the extent of fundamentally altering or threatening existing power relations; rather, it has been appropriated, initiated, and directed by dominant discourse and those existing power relations.

These positions sought to account for changes that were perceived to be happening as industrialism morphed into post-industrialism and have been developed considerably in more recent years. In addition to the speculative turn discussed above, the ways in which media are perceived and assessed have been developed in relation to 'media ecology', most significantly by Matthew Fuller (2007). But such contemporary approaches also have their precursors.

The notion of an economy of complex interrelated factors engaged in a constant process of co-evolution was set out by the Santa Fe group of researchers as they searched for alternative explanations for unpredictable economic behaviour. The idea of unpredictability that they explored is central here. An economy is not an enclosed totality that can be observed as a static environment in equilibrium suited to deductive reasoning, but rather it is a constantly shifting noisy ecosystem of co-evolving relative parts. The idea of an ecosystem fits well with the aim of this book and it is adopted here to account for the complexity of digital space. Any analysis of such a system requires an inductive approach of the kind proposed by Francis Bacon, as quoted by Marshall McLuhan above, and more recently by researchers from the Santa Fe group.

Two of the Santa Fe group's key members were Brian Arthur and John Holland, an economics professor and a computer scientist respectively (Waldrop 1992). Together they would reinterpret the rules that had dominated economic thinking for decades, if not centuries. For them, economics operated as a complex system that was always in process and which echoed complex pattern formation in natural ecosystems. However, economics was complicated even further for them by the addition of human agents and the unpredictable decisions they made (Arthur 1999). The unpredictability of humans is an important factor in the argument being developed in this book. But this unpredictability did not deter them (or us). They asked the questions: 'How do you think about an economy that is always evolving, that never settles down to an equilibrium? If you do computer experimentation in economics, how would that work?' (Waldrop 1992: 250). This relates once more to the question of speculative realism discussed above and the possibility of a mathematical model of life that resides in relation to, but separate from, lived experience. Which is real is the question that Meillassoux grappled with. It is also the one that Hervé Fischer raises in his *Digital Shock* (2006). For our part, the question is largely unnecessary as the co-evolution that takes place in John Holland's computer simulations, is regarded as occurring not just within those systems but between systems and the world they try and model. Thus, it fits with idea of a back and forth movement between the digital and the analogue as discussed in Chapter 3, in relation to Brian Massumi.

The Santa Fe group debated the idea of bounded rationality, the idea that decision-making was a rational process that could work appropriately within the constraints of its condition: the amount and kind of information available, and the time constraints, etc. Economics had long worked on the assumption that the decision-making process was rational, and lacking mystery. But Arthur and his colleagues were less sure. They asked the question: where do you set the dial on rationality? The answer they arrived at was, you don't. According to Arthur:

> You'd pull a John Holland on it. You'd just model all the agents as classifier systems, or as neural nets, or as some other form of adaptive learning system, and then allow the dial to vary as the agents learn from experience. So all the agents could start off as perfectly stupid. That is, they would just make random, blundering decisions. But they would get smarter as they reacted to one another.
>
> (Arthur, cited in Waldrop 1992: 251)

The approach of the Santa Fe group was described by Waldrop in the following way:

> Instead of emphasizing decreasing returns, static equilibrium and perfect rationality, as in the neo classical view, the Santa Fe team would emphasize increasing returns, bounded rationality, and the dynamics of evolution and learning. Instead of basing their theory on assumptions that were mathematically convenient, they would try to make models that were psychologically realistic. Instead of viewing the economy as some kind of Newtonian machine, they would see it as something organic, adaptive, surprising and alive. Instead of talking about the world as if it were a static thing buried deep in the frozen regime, as Chris Langton might have put it, they would learn to think about the world as a dynamic, ever-changing system poised at the edge of chaos.
>
> (Waldrop 1992: 253)

Holland had been instrumental in showing the group that the environment in which economics took place was not stationary. But that was not to say that it did not demonstrate forms of evolutionary patterning that could be tracked and understood – to a degree. This describes pretty well the approach being developed here: a sonic economy is a methodology that allows for critical engagement with, and the identification of, the heterogeneous apparatus of politics and power that is operational at any given time in a complex system. It is an economy that recognizes difference as a positive and often unpredictable force, and that instead of trading in exchangeable generalities, identifies repeating singularities that cannot be reduced to equivalence.[3]

As industrialism morphs into an information society in which the dispersion of discrete elements is present but not always clearly visible, or predictable (not visible as a coherent whole enclosed in perceptible space, but rather as a fluid, mobile, and therefore temporal experience), an appropriate dynamic, mobile mode of analysis is required: one not tethered to representation, that can accommodate an almost perpetually shifting ground that is akin to Rancière's curiosity shop, discussed in Chapter 3.

Bestowing materiality on that which cannot be actualized visually, while recognizing the influence of Marx's 'Theses on Feuerbach' and Bataille's concept of 'Base Materialism', and Greg Hainge's clarification of the concept of noise – insofar as it is a materiality that resides in the day-to-day rhythms of the urban milieu – is not an act intended to represent a new kind of truth that heralds particular ends or beginnings. Instead it poses the dilemma that when you *look*, you can't always see the connection between people/institutions/events/phenomena that common sense tells us are separate and unrelated, or at least belonging to different realms and operating according to different rules; but if you *listen*, you can discern the polyphonic political vibrations.

Listening then becomes an analogical strategy for political engagement, a means of making connections across time and space between people/institutions/events/phenomena. It is sometimes but not always literal insofar as it proposes the strategy be employed to demonstrate the interrelationship of factors that do not necessarily or exclusively reveal themselves on an audible sonic register.

As an approach it is necessary because, as has been stated by the online music and media distribution organization, Slow To Speak:

> Life in its truest form is a kind of music; characterized by propulsive syncopated rhythms, polyphonic ensemble playing, varying degrees of improvisation, and often deliberate distortions of pitch and timbre. Implied is

> a series of balancing acts. It must always be disciplined – but never driven – by formulas, agendas, or sheet music. It must always be pushing outward, forward, upward – and therefore, inevitably, against complacency. Slow To Speak exists to run parallel with life's music. Slow To Speak aims to offer its own distinctive melodies, rhythms without pretence and the entrapment of modern life. It is a groove that ignites passion – hot beads of sexual excitement, a groove that is gravitational – creating the sweet funk of action and influence. Slow To Speak is raw, convicting and at times uncomfortable – setting itself apart as a media outlet subsisting on truth and its convictions . . .
>
> *slow to speak* is taken from a Bible verse found in the book of James, 1:19. It reads, 'be quick to listen, slow to speak and slow to anger.' This is the approach we are taking to compiling and sequencing music. It is one that begets genuine appreciation for the work, its presentation and the audience.[4]

The term 'sonic economy' operates in accordance with this statement. It is a mode of thought where multiple aspects of production, communication and exchange are assigned and/or assume interrelated value, duration, and speed/tempo. As an approach, it imbues notions of rhythm and harmony with a materiality not present in much contemporary analysis. Hence, political economy as an approach to understanding the complexities of the contemporary technologically mediated environment is reconfigured using approaches that have hitherto resided in the apocryphal realm of music or sound more generally.

The term 'political economy' is used here to describe an amalgam of factors relating to the systematic management, production and distribution of goods and services/information/data, and both the formal and informal political environment in which that process takes place. Too often, previous attempts to understand the complexity of such arrangements have done so in a manner characterized by neat periodizations with attendant beginnings and ends, as well as clearly marked lines of demarcation between specific spheres of activity.[5] The largely static nature of such analyses, however, relying as they so often do on representational certainty and dialectical rationality, cannot adequately account for the seemingly relentless drive towards the ubiquity of the now not-so-new and converging communication technologies or their coming into being.

The emergence and continued development of such technologies form a story that is not easily told. Formal narrative structures that employ a beginning, middle and end are unable to account for the complexity of this particular tale.[6] Dialectical and representational approaches that ignore such complexities, serve to fix and represent spatio-temporal phenomena according to a particular logic. They do so with a claim to authenticity that endures until an acceptable challenge is instituted and sanctioned. History and progress are set out in a linear succession of periods or blocks that subsume all in their continuous path. Heidegger called it ontic knowledge – a kind of ordering principle that could explain the world and our place in it to a degree – a primarily visual degree that had come to dominate Western thinking since the time of Plato. But what of the invisible, the things that exist in the folds and shadows, the things that fail to conform to conventional explanation? Are they simply not there, or, more worryingly, to be classified as unimportant or as statistical anomalies? And, finally, where does it leave sound? What about phenomena that reveal themselves to the ear but not to the eye? In the multimedia, technologically mediated environment that we currently inhabit, the importance of sound as well as other sense stimuli needs to be restated.[7]

But the call for a sonic method is not simply promoting the case for an alternative – one mediated via exposure to the forces of negation to produce a concrete new implacable method, but rather is intended to set method free in the digital sense of multiple possibilities for connection across space and time. It does not, for instance, call for a dialectic struggle between formal political/economic and creative sectors, as was the case recently with the Demos Report on Cultural Diplomacy mentioned above.[8] That report advocated a modified form of economic and political imperialism that embraced and took seriously the power of contemporary British cultural products. In this particular case, highlighting the diplomatic potential of the band Razorlight as a means of advancing Britain's standing on the global stage. Simply putting music into the equation is not the same as a sonic economy.

Finally, the term 'sonic' with its connotations of speed and movement, is conjoined with 'economy' because of the already stated interest in material exchange value, and reciprocity, which, while dealing with issues of ontology and or metaphysics, is, as well as being a philosophy of historical movement and connectivity, also an economy of immanent relations between singularities with lines of flight as trade routes and feedback loops forming sonically actualized activity that operates in relation to the management, mismanagement or outright failure to manage, of resources, physical and otherwise, in a manner that goes beyond the formally political or economic without leaving them behind. It therefore considers factors not accounted for in the strict parameters of political economy.

The sounds that resonate in certain specific political locales or milieux, as will be explored below, can tell us something significant about a particular place and time, and arranging political, economic, technological and cultural factors, into 'harmonious scores' will be beneficial. Music/sound/noise should therefore be regarded as important in and of themselves and also as interrelated frameworks for understanding complex historical machinations where timing replaces time in a space where diverse statements are *arranged* into discourses or agendas.

The notion of an agenda is posited here as an arrangement of often disparate elements, or an interplay of harmonies and melodies, across time and space. They operate sonically, picking up mood, rhythm, timbre and tempo, and then sit out a few bars in silence as other momentarily more apposite factors take their place. An agenda is not a hierarchy but a bringing to order; not always logical or possessed of clear motive, but always demonstrative of a process of power in operation.[9] Accordingly, practically distinct spheres of life such as politics, philosophy, art and science, can and do exist within and between agendas, and are not subsumed under a universalizing meta-narrative or ideology that can be readily identified or represented. They exist in their own right, as political economy approaches have shown us, but they also exist in relation to one another. Like distinct items on an agenda or aspects of a musical arrangement, they play off each other. For this reason we need to be in a constant state of readiness in our assessment of which line is resonating with which other and when.

A sonic examination of such agendas can assist us in making sense of the complexities of contemporary technological environments when simply looking may not be enough.[10] The task of arranging the sometimes invisible, which finds its theoretical ground in Heidegger and later in Deleuze and

Guattari as will be shown below, draws together two aspects of Michel Foucault's work, namely *discourse* and *blinding luminosity*.

For Foucault: '[T]he term discourse can be defined as the group of statements that belong to a single system of formation; thus I shall be able to speak of clinical discourse, economic discourse, the discourse of natural history, psychiatric discourse'[11] (Foucault 2004a: 121).

Laclau and Mouffe describe a similar category that they call *articulation*. In a manner that directly echoes Foucault, they say: 'We will call *articulation* any practice establishing a relation among elements such that their identity is modified as a result of the articulatory practice. The structured totality resulting from the articulatory practice, we will call discourse' (Laclau and Mouffe 2001: 104). Thus, it might be asked: is it now possible to speak of a discourse on technology? One that is multiple and dispersed, that cannot be seen but which can be heard. In this respect I draw again on Laclau and Mouffe who have pointed out Foucault's rejection of four hypotheses in relation to principles around which discursive formations might cohere. They are: '[R]eference to the same object, a common style in the production of statements, constancy of the concepts, and reference to a common theme. Instead he makes dispersion itself the principle of unity, insofar as it is governed by rules of formation, by the complex conditions of existence of the dispersed statements' (Laclau and Mouffe 2001:105).

Such dispersal renders certain phenomena invisible. 'Technology', for instance, is not a specific object that can be seen; it is present and spoken of in a number of forms that in turn are not stylistically uniform. Many concepts are drawn on: historical, scientific, artistic, and philosophical. In terms of themes, again, these are multiple: political, socio-cultural, economic, human/post-human, etc. The process of bringing them together and making connections is what is being described here as sonic. That the discourse can be identified in different forms in different dispersed locations is the very point that demonstrates its significance as a category; but it also makes it difficult to identify specific discourses using conventional means. This difficulty is raised by James Faubion in his Introduction to Foucault's *Death and the Labyrinth* (2004c) where he highlights the way in which Raymond Roussel's literary cosmos is shown to operate as one of thresholds and parentheses where elements (or discourses) get closed off but then suddenly open up in relation to other elements to create a kind of unity. Once more, however, it is a unity that is not always clearly visible: 'It is a place of relentless spectacle, of sheer visibility, but of a luminosity so intense that it can be disorientating, even blinding. It is thus a place in which what is most fully exposed has perhaps the best chance of remaining secret'(Foucault 2004c: vii).[12]

Hence the trade and communication between discrete elements or discourses needs to be established using a sonic methodology that highlights the extent to which multiple discursive practices, moving at varying speeds or tempo, periodically cohere to form a 'refrain' in the sense that Deleuze and Guattari use the term. Accordingly, phenomena such as technological development or technology as discourse, can be said to be set to a particular score – synchronized and harmonic.

Reference to Deleuze and Guattari in relation to this matter assists us in a questioning of Heidegger's essentialism and the ultimate dialectical choice between modern technology and a return to a kind of reconceptualization of technology that embraces poiesis as set out in his work, *The Question Concerning Technology* (1977). Heidegger's phenomenology/ontology of Being is useful in moving us beyond Platonic knowledge based on the primacy of visual stimuli, stating as it does that just because it cannot be seen does not mean it is not there or that it does not possess materiality. But perhaps of more use in this respect is the concept of the refrain – not as metaphor but as operational mode of analysis making multiple connections within a rhizomatic 'world wide web of significance'.

'Of the Refrain'[13] describes sonic milieu components as sounds that perform a specific function in any given territory, as markers of familiarity and danger. They are the environmental noises that are not necessarily coherent or ordered. The refrain, on the other hand, resolves to do something different: it works not as an opposition but like the resistance that Greg Hainge described above. It can be organized not just sonically but as the rhythms and patterns of everyday life that Lefebvre describes. The refrain calms the chaos – settles things down – resolves anxiety, but as music that exists *in* noise rather being its opposite, it can also assume singular difference that can deterritorialize the political calm, as was the case in both Coventry and Detroit that will be discussed in Chapter 5. It can disrupt the rhythm creating a momentary energy field that distorts the dominant order of things as specific spheres or milieux in Roussellian parentheses open up, creating thresholds for movement and connection with the outside.

> Thus philosophy, art, and science come into relations of mutual resonance and exchange, but always for internal reasons. The way they impinge on one another depends on their own evolution. So in this sense we really have to see philosophy, art, and science as sorts of separate melodic lines in constant interplay with one another. With philosophy having in this no reflective pseudoprimacy nor, equally, any creative inferiority. Creating concepts is no less difficult than creating new visual or aural combinations, or creating scientific functions.[14]

The concept of the sonic then operates on two levels here: as a philosophy of non- representation that embraces movement and fluidity, and at the level of music/sound/noise as significant sites for investigation. In doing so, it echoes Jacques Attali's *Noise: The Political Economy of Music* insofar as it concentrates its attention on that which is often marginalized in relation to more 'legitimate' political and economic concerns. Attali has stated: 'Today, our sight has dimmed; it no longer sees our future, having constructed a present made of abstraction, nonsense, and silence. Now we must learn to judge a society more by its sounds, by its art, and by its festivals, than by its statistics' (Attali 2006: 3).

As useful as Attali's work is, it still operates, as the title suggests, at the level of traditional political economy. Its principal claim being that an analysis of music from a political and economic perspective will allow us to identify the future of the wider political economy, and, in our case, the broader digital landscape. And Attali is not alone in proclaiming the predictive qualities of sound and sound-related technologies. Jonathan Sterne has said:

> Although it is a ubiquitous and banal technology, the MP3 offers an inviting point of entry into the interconnected histories of sound and communication in the twentieth century. To access the format's historical meaning, we need to construct a new genealogy for contemporary digital media culture. Many of the changes that critics mark as particularly salient aspects of contemporary digital or 'new' media happened in audio before they surfaced in visual media. As Frances Dyson writes, digital media encapsulate 'an accumulation of the auditive technologies of the past.' The historical resonance of audio

> can be extended across the various registers of new media, from their sensual dimensions in both the auditory and visual domains, to their treatment of subjects, to their technical structure and industrial form.
>
> (Sterne 2012b)

The point of a sonic approach is not to necessarily privilege sound, art, or festivals, but to place them in relation to other phenomena in a manner where emphasis (on value/duration/speed/tempo) is constantly shifting. It is necessary then to apply sonic thinking to the statistics and non-auditory phenomena as well as to festivals. To do so will require that Jacques Attali's political economy approach be augmented. This can be achieved by introducing what Pauline Oliveros calls 'Deep Listening'. It is a process that, like the one proposed above by Sara Maitland, unifies the senses in acoustic space and which facilitates an awareness of the present and the immediate as well as trajectories and sequences in relation to other phenomena across what she calls the space/time continuum. Oliveros goes on to make the further very important point that Deep Listening might be usefully applied to aspects other than the audible; to urban environments and broader political economies as well as to sound. In doing so, connections can be made between dispersed phenomena – without privileging one over another – across time and space that might not be visible to the naked eye: 'The level of awareness of soundscape brought about by Deep Listening can lead to the possibility of shaping the sound of technology and of urban environments. Deep Listening designers, engineers and city planners could enhance the quality of life as well as sound artists, composers and musicians.'[15] Such an approach can tell us something about politics, power and resistance, with music and sound existing not simply as metaphor but as significant nodal points in a complex contrivance of interconnected factors, each inhabiting an expanded, wider and deeper, political terrain where we must listen to dispersion.[16] If looking confirms abstraction, then listening establishes connections. It is this musical principle – the arrangement of resonant phenomena – that is being proposed here and which constitutes a means of sonic critique that supplements all the other senses. 'Life in its truest form is a kind of music', made up of immanent and material relationships between dispersed phenomena.

Notes

1 See Winner, L. (1993). Also see his 'Do Artifacts Have Politics?'

2 This can be read against Harman's rejection of relativism and speculative realism's rejection of correlationism as discussed above.

3 See http://foucaultblog.wordpress.com/2007/04/01/what-is-the-dispositif/ [Accessed 10 Sept. 2014].

4 See http://slowtospeak.net/originalabout.html [Accessed 10 Sept. 2014].

5 Habermas draws a distinction between what he calls the 'system' and the 'lifeworld' as independent spheres of activity. See http://www.csudh.edu/dearhabermas/habfouc02.htm [Accessed 10 Sept. 2014].

6 The problems associated with the strict assignment of beginnings and ends are discussed by Jean-François Lyotard in *Soundproof Room: Malraux's Anti-Aesthetics* (Stanford, CA: Stanford University Press, 2001), pp. 2–7 and also by Gilles Deleuze in *Negotiations* (New York: Columbia University Press, 1995), pp. 121–2.

7 In doing so, it is useful to reflect on McLuhan's concept of Acoustic Space:

> Acoustic space is a complete contrast to visual space in all of its properties, which explains the wide refusal to adopt the new form. Visual space, created by intensifying and separating that sense from the interplay with the others, is an infinite container, linear and continuous, homogenous and uniform. Acoustic space, always penetrated by tactility and other senses, is spherical, discontinuous, non-homogenous, resonant, and dynamic. Visual space is structured as a static, abstract figure minus a ground; acoustic space is a flux in which figure and ground rub against and transform each other.
>
> (McLuhan and McLuhan 2007: 33)

8 The relationship between formal political/economic factors and the 'creative sector' can and should be related to the distinction, mentioned above, that Habermas draws between the 'system' and the 'lifeworld'. See: http://www.demos.co.uk/publications/culturaldiplomacy [Accessed 10 Sept. 2014].

9 In previous research for his PhD in 2003, the author found the term 'The Technology Agenda' has been used to describe how disparate statements in the Foucauldian sense of that term, across a range of mediated forms and emanating from distinct spheres of life – politics, economics, as well as the media and the arts – could be somehow sutured together to give us a better understanding of the way in which discourses work to normalize technological life, or in this particular case, the use and adoption of developing network technologies.

10 In the same way that listening to the stars can reveal something deeper than just observing them; listening to political, economic, technological and cultural shifts will reveal something significant about the movement and intensity of power and resistance. See http://bbc.co.uk/1/hi/sci/tech/7687286.stm [Accessed 10 Sept. 2014].

See also, singing genetic code: http://www.bbc.co.uk/news/10581179 [Accessed 10 Sept. 2014]; http://www.bbc.co.uk/news/science-environment-19647038 [Accessed 10 Sept. 2014]; http://www.bbc.co.uk/news/science-environment-18449939 [Accessed 10 Sept. 2014].

11 Michel Foucault, *The Archaeology of Knowledge* (London: Routledge, 2004).

12 Michel Foucault, *Death and the Labyrinth* (London: Continuum, 2004), p. vii.

13 In Gilles Deleuze and Félix Guattari, *A Thousand Plateaus*, trans. B. Massumi (Minneapolis, MN: University Minnesota Press, 1987).

14 See Deleuze (1995), p. 125.

15 Pauline Oliveros, *Deep Listening* (New York: iUniverse, 2005), p. xxv.

16 Dispersion is a concept that applies readily to music, with technology allowing for the movement and reconnection of musical fragments across time and space in the form of samples and the multiple generic arrangements so common in electronic music. Listening to such dispersed fragments implies the creation of a narrative that when it is itself listened to deeply, goes beyond music to allow connections to be made with all other aspect of life so that a unity of narratives without specific beginnings or ends can be formed. This relates to Foucault's work in *Death and the Labyrinth*, and also to Lyotard's *The Soundproof Room* (2001).

5

Motor Cities

It is not uncommon for parallels to be drawn between the cities of Coventry and Detroit. The familiar themes of motor manufacturing, racial division and music have led commentators to describe the rise of Two-Tone Records in Coventry by referring to the city as 'something like the UK's Detroit'.[1] And a recent radio documentary in Detroit reciprocated the comparison.[2] Such comparisons, based as they are on commonly assumed concepts of identity and difference, will be addressed here.

Coventry is situated 52.4081° N, 1.5106° W, Detroit is situated 42.3314° N, 83.0458° W. They are 3,668 miles apart yet often linked conceptually due to the recurring motifs of cars, industrial decline, ethnic diversity, and urban decay. An examination of these resemblances raises the question of the qualitative nature of place and the, to a degree, quantitative, or correlative nature of space as discussed in Chapter 1. It also raises the related question of difference. How are the two places different, and can they ever be the same? What is the difference between Coventry and Detroit? When are they different and when are they the same, and why does it matter? It may be that the only similarity/resemblance is that they are both different – not equivalent or exchangeable in a representational economy, but different in a sonic, non-representational economy, and in the way that Gilles Deleuze described difference above. Resemblance and generality, as has been discussed, are of only limited use. What are more useful are the intensities that repeat.

The eradication of 'difference between' as initiated by Deleuze, assists us in addressing the problematic nature of resemblance and equivalence and calls for new economic models that can account for the uncertainty of what something is when it can take multiple forms and adopt superpositions in digital space and time. This chapter will examine in detail this uncertainty. Specifically, it will examine the relationship, similarities, and differences between Coventry and Detroit – from the late 1970s to the digital turn of the 1990s – and assess the extent to which the exposition of significant events in the two cities and their interrelatedness across space and time can inform a better understanding of the shift from modern analogue industrialism to a technologically mediated digital age. In doing so, it will implicitly draw on the spatial categories of Foucault as set out by Deleuze in the chapter in his (1999) book, *Foucault*, entitled 'A New Archivist: Collateral'; *Correlative* and *Associate* as described above. It will also operate as an accompanying soundtrack to the decline of two industrial giants. As Coventry and Detroit experienced economic catastrophe, what did the sounds that came out of these cities tell us about the digital turn? And by examining this question from two specific spatial viewpoints, how can we support the argument that visual logic prefaced on redundant notions of difference and repetition, are incapable of accounting for this turn or the time period that it preceded?

Coventry and Detroit demonstrate specific spatialized contexts out of which musical movements developed to disrupt and resist the dominant political order of the day. They are also both examples of places where music external to the territories in question entered the milieu, colliding with rising discontent and political unrest. In the case of Coventry, it was Ska music escaping the milieu/territory of Kingston Jamaica as a line of flight – slipping its moorings literally as ships brought immigrants to the UK – bound to make contact with a disenfranchised and largely unemployed section of Coventry's community. In the case of Detroit, it was the computerized music of Kraftwerk that slipped its moorings in Germany to connect with similarly disenfranchised groups in Detroit. Both movements took (and in the case of Detroit continue to take) milieu components and organized them into refrains that served to create new territories and communities with particular political agendas. With music at their core, both were able to disrupt the dominant refrain, serving to deterritorialize and challenge the prevailing political wisdom.

While channelling these deterritorializing forces, it is important not to make claims in relation to incidences of cause and effect, beginnings or ends: economic decline causing racial tension, unemployment, political unrest and in turn a fertile arena for creativity, for instance. Instead it is essential to search for confluences, moments of interaction and collision. Rather than simply *looking* for connections, we are going to listen for them. In doing so, the aim is to challenge assumptions about cause and effect in both examples and by virtue of embracing the sonic as affect not effect, allow for the positing of a new kind of political philosophy that is in tune with the contemporary moment. Investigating events within an acoustic environment allows us to extricate ourselves from closed systems of thought. Methodologically, then, we will be able to make precise statements without recourse to universal laws.

The overall intention is to reconfigure political economy without recourse to dialectical thinking. The digital age did not simply replace the industrial. It therefore requires rethinking in relation to non-linear methods of acoustic thinking appropriate to this age of technology, such as the 'deep listening' of Pauline Oliveras (2005), who urges that we listen not only to sound but to the political economy of the urban environment. A political economy approach might reveal something interesting about specific dynamic factors in each case, but when augmented by 'deep listening' a whole new horizon or 'time–space continuum' opens up in which expanded communities of interest (connections) can be identified. The connection between Coventry and Detroit will be explored at this level.

An analysis of the economic decline of Coventry and Detroit will be set against the more general economic and political malaise in the UK, and in the US that was linked to the collapse of the car manufacturing industry and the musical phenomenon that accompanied it/preceded it as fanfare.

To address the failing of approaches hitherto adopted, this chapter will be using a methodology that is being called sonic economy. Echoing the critical methodology of political economy that for so long helped to reconstitute and bring to light the hidden nuances of power, politics and the socio-cultural networking of an industrial age, it will foreground the importance of music specifically, and sonic phenomena more generally, as significant factors in understanding the unfolding events that together have constituted the digital age. Drawing on aspects of political economy, philosophy and cultural geography this chapter examines how the urban spaces in question have become reconfigured to form post-industrial landscapes with intricate networks of connection and exchange that defy explanation in traditional terms.

Rethinking political economy in this way means pursuing a non-representational enquiry that draws on sonic rather than solely visual data to develop a more reliable account of the decline of traditional economic sectors and the so-called technological turn.

The importance of place: Coventry

> *Coventry's sudden rise in motor manufacturing – earning it the title of motor city or 'Britain's Detroit' due to the large concentration of car production plants – saw the city centre being redeveloped to accommodate an increase of moving traffic.*
>
> *However the creation of a concrete collar around it for a ring road has perhaps been to the detriment of communities outside of it and for businesses inside it, which have struggled to expand.*[3]

In the period leading up to the Second World War, Coventry was the fastest-growing urban centre in Britain. And this continued after the war as the city emerged from the rubble to become a central part of the new Labour government's vision of a new Britain based on manufacturing output, technological advance, and urban development. If we accept the claim that spaces speak, and the related call that we should listen, then Coventry could be heard to be speaking with a renewed confidence (Blesser and Salter 2007). It boasted the first pedestrianized shopping centre in Europe – a space where having left their homes and parked the car (and Coventry had a higher rate of car and home ownership than any other industrial city), citizens of this shiny new city could celebrate their newly acquired prosperity. Its manufacturing and engineering workers enjoyed unprecedented wages, and immigrant labour keen to take advantage of the economic upturn, came from all areas of the British Isles, Europe and the Commonwealth to share in its prosperity.

> Its new cathedral consecrated in 1962 and dedicated to the cause of international peace and reconciliation, seemed to symbolise a prosperous and outward-looking future for a city. But the spiral of history was about to take another cruel twist. When the clouds of the recession, fuelled by an oil crisis, began to gather at the beginning of the 1970s Coventry had neither the diversified industrial base nor the firmly-rooted corporate spirit to blow them away.[4]

The economic success did not last. As the 1980s approached, the machine-like efficiency that characterized the city, and which provided the predictable and steady rhythmic backing to its fortunes was beginning to miss a beat and to run out of time. With the collapse of the Alfred Herbert Company, once the world's biggest machine tool firms, and the subsequent machinery sales that were held in the factory where thousands had previously worked, Coventry began to dance to a different tune.

More factory closures followed, and the failure to invest in its civic infrastructure, and its by now ailing industry, brought increasing unemployment rates. Coventry was now a bleak space. The concrete that had once characterized its rise from the ashes as a modern industrial giant, and which once shouted proudly to the rest of the world, now whispered an eerie ghostly lament.

The population of the city peaked at 330,000 in the early 1970s, but by the turn of that decade the decline had begun. The story of this decline has been well documented but by way of a summary:

> In 1974, 115,00 people were employed in manufacturing in the city ('boom town'). By 1982, the figure was 62,400, a fall of 46 per cent ('ghost town'). 40,000 of the 53,000 job losses occurred after 1978 as the de-industrialization of the early 1980s took hold (Clark 1989: 53). This decline hit Coventry's core motor manufacturing plants particularly hard.[5]

During the period 1978–1982 (*The Year of Technology*), a chaotic noise had begun to form in Coventry. This chaotic noise was audible not only in the music that was forming, but also between that music and wider social, economic and political concerns. The sound was one of decay as the concrete on which the economic boom had been built started to disintegrate, both metaphorically and literally. In Coventry, the 'concrete collar' was choking the city to death. But it fought back, kicking and screaming, and those screams would echo in distant places (Detroit) as the energy was dispersed and then re-amplified. The noise was the Two-Tone sound of Ska. It was a sound that combined the energies of what had previously been divided into various generic categories of either black or white music, to produce a reinvigorated kind of political engagement.

As the economic downturn gathered pace, this strange noise that was coming out of Coventry came to be audible. It was a noise made by the band, The Automatics. Of their later released album, *Dawn of a New Era*, Dave Thompson has written:

> Essentially, they were the sound of a group of players with various musical inspirations, trying to bring them all to the fore at once, while clinging to the none-too-focused notion that they're playing reggae. Rockabilly riffs vie with big rock lead guitar flourishes; of the songs that would, in later years become a national obsession, few are recognizable.
>
> (Thompson 2004: 21)

The Automatics emerged from a Coventry music scene that was richly diverse. The band's founder Jerry Dammers had been playing keyboards with groups such as the Sissy Stone Soul Band and the Ray King Soul Band through the 1970s, but when punk rock exploded in 1976, his energies were refocused. His soul legacy would merge with punk, and its newly realized ally reggae, to form a new sound that was more than just the sum of its parts. From its inception, punk had formed an alliance with reggae. '[W]hen punk's own mecca – the Roxy Club in Covent Garden – first launched in 1977, it was chance, not design that brought Reggae onto the dancefloor' (Thompson 2004: 19). This chance happening fits perfectly with the chaotic themes of this book. Don Letts, the resident DJ, later reflected that there had been no punk records to play, so he simply played what he liked, and the DIY aesthetic of much of the reggae he played resonated with the punk ethos. This collision found expression in the Clash's cover of Junior Murvin's 'Police and Thieves', and in the repertoire of Alternative TV and Patrick Fitzgerald. But it was not until The Specials that it really found its expression culminating in 1981 with 'Ghost Town'. For many, it defined an era: see http://www.bbc.co.uk/news/magazine-13780074 [Accessed 10 Sept. 2014].

The Specials had grown out of The Automatics. They had changed their name to The Coventry Automatics to assert their descent and to avoid confusion with another band of the same name. They would later change it again to the The Special AKA The Automatics, and then just to The Special AKA, eventually settling on the name The Specials with the release of their second album. When the band split in 1981, and with key members going on to join the Fun Boy Three, the remaining members returned to The Special AKA. This constant changing of identity was significant in so far as it allowed the group to always be opening out to new opportunities and challenges, always something other than what people thought they were: the sound of Britain in chaos. They were a positive political force, a difference that repeats with an identity that resisted representation, and which can be contrasted with the generalized resemblance of the Specials that reformed, minus Jerry Dammers, in 2008.

The year after 'Ghost Town' had seemingly captured the bleak reality of a city and a country in steep industrial decline, was the officially designated *Year of Technology*, and with The Specials now returned to The Special AKA, the question being asked here is whether this was a significant turning point. It would certainly signal the seminal moment in the career of the group. From this point on, their popularity would begin to wane, eventually giving way to a new kind of music from, the US and more specifically from Detroit, that went on to find its first officially sanctioned home in the now legendary Eclipse club in Coventry.[6] The transition was the sound of the world coming to terms with the new post-industrial order, and it had a particular register. Listening to it tells a story not just of structural decline but of the positive forces of voices straining to be heard.[7]

In Coventry, the punk/reggae synthesis, or collision, was the result of a union of like-minded souls, or events as they might better be called. The collision occurred in a particular spatial context as people, objects, money, materials, emotions, configured, dispersed and reconfigured as part of a sonic economy, as the industrial practices that formed around large-scale manufacturing and the fusion of related social forces and population flows were discontinued and cleared, to be superseded by new practices that were beginning to emerge. Such collisions were contiguous (Shaviro 2012) in spatio-temporal terms, common to time and place: in this case, Coventry in the 1980s. But they also occurred in a noncontiguous space, demonstrating common defining characteristics 'inherited from one another' and in so doing demonstrate how coherence and resonance can be identified in a non-representational manner. It is in this way that Coventry and Detroit cohered – at the quantum level, as echoes of each other (Shaviro 2012).

This was not the first time politics and music had collided, and as a narrative without beginning or end, Coventry found itself caught in a 'moment'.

> The difference was, those past insurrections had challenged the Powers That Be, the governments of the day. The uprising that began in 1978, with the first brickbats of discontent on the streets of Britain's inner cities, and climaxed in 1981 with a summer of unprecedented riot, went beyond those petty aims, to challenge institution itself, to take on an engrained way of thinking, a physical way of life.
>
> (Thompson 2004)

This idea of a physical way of life is interesting in relation to many of the ideas and theories set out here. In this instance, or on this occasion, music, protest, politics, economics etc, were not separated but cohered to form a physical, objective world of embodied energy. It is not necessary to claim this moment was the first of its kind, as originary, but it is important to recognize its intensity and status as event, or as repetition. To do so invokes McLuhan's acoustic space, the seven tenets of non-representational theory and the collapse of the distinction between the system and the lifeworld, as music took its place among other significant phenomena, not as oppositional but as expressive force in the way that Hainge described in Chapter 4 (Hainge 2013: 16). It was part of an affective, immersed and non-hierarchical discursive practice where all elements that together constitute a place, a society or an economy, can be deemed significant. The story of a band's formation and the development of a musical style say as much about the world as formal political economy. It was not a case of music that addressed political issues, but the sound of a city straining at the limits of its endurance as the analogue wheels of industry slipped out of gear and began to slow almost to a stop – without ever stopping entirely. This was the story of Coventry *and* Detroit.

The importance of place: Detroit

Detroit endured an even more profound decline than Coventry. Its fortunes, like Coventry's, were linked to the fortunes of its industrial base. During the Second World War, it became its country's main producer of arms, and through this period saw a massive influx of workers, and the city's population peaked in 1953 at 1.5 million. During this time it also witnessed a bifurcation on the grounds of race as white workers migrated to the suburbs while black workers took up residence in the city. Ever since, the city has experienced a massive decline, and this has been well documented and debated.[8] Through the race riots of the 1960s, and the radical politics of the Black and White Panthers respectively, the city endured a monumental downturn, and now has a population of around 700,000. This downturn, as was the case in Coventry, was inextricably linked to the motor industry.[9]

> The beat of the bongo drum, the rattle and hum of factory machinery, the boisterous rhythms of Martha Reeves and the Vandellas' song ['Dancing in the Street']: these sounds, heard in isolation, might seem innocuous enough. But considered in context, in the charge of their historical moment, the sounds acquire new meaning.
>
> (Smith 1999: 5)

In her *Dancing in the Street: Motown and the Cultural Politics of Detroit*, Suzanne E. Smith presents a case for the specific role of place (and therefore space in terms of the argument presented above) in any analysis of the production of culture, in this case music and the Motown phenomenon in Detroit. It is an argument that would later be echoed by Kilian Hanf in relation to techno:

> A machine has no soul. Yet its execution of human ideas can be conceived as the transfer of the soul through the machine. At least, in the Detroit of the 1980's, a number of black musicians saw it that way. Detroit techno, the electronic music they created during that time, was exactly based on this philosophy. The machine – the synthesizer – reflected the soul – the condition of a decayed and dangerous, futuristic and fertile city and the people living in it.
>
> (Kilian Hanf 2010: 3)

These quotes are useful reference points to reflect on in relation to the task at hand here that, while not seeking to add to the volumes already dedicated to Motown, or techno for that matter, is concerned with the specific characteristics of Detroit. For both Smith and Kilian Hanf, you cannot uncouple the sound from the place.

According to Smith, many writers, music fans, and scholars have argued that Detroit in *not* critical to understanding the Motown phenomenon. For them, Motown could have happened anywhere, or at least in any city with a large African American population. It could, they say, have happened in Chicago, New York, Pittsburgh, or Cleveland.

Smith argues that these arguments tend to emphasize individual ambition rather than community life, urban geography, economic structures, or race relations as factors in Motown's rise to the top of popular music. But here they are taken to be examples of the emphasis on generalities and resemblances, which can easily be exchanged in order to account for situations in places that are judged to share the same characteristics or belong to the same genre or set. What is of greater concern here are the energetic and dynamic differences of each place where neither individual ambition, not social factors alone produce outcomes, but rather complex and intertwining relationships that produce distinct singularities.

This is significant in so far as it places the individual, or individual ambition, within a network of other significant factors. Motown, Smith claims, takes its place in and amongst other significant phenomena/events that emerged from 'black Detroit' (Smith 1999: 8). In order to support the argument, she invokes the theoretical insights of Raymond Williams in relation to cultural formation. Williams argued for the interrelationship between artistic forms and social location – the way in which a phenomenon forms is as important as the phenomenon itself. This argument is further developed within the context of this study where phenomena are approached from the perspective that they are in a constant state of formation (in-formation, holding steady for a moment before dispersing and reforming – as noise as Hainge would have it). 'Motown did not stand apart from its social circumstances but was the product of a complex set of historical forces' (Smith 1999: 10).

The racial mix, the civil rights movement, the industrial structures and work practices; the production line, the emergence of a specifically 'black capitalism' were all significant for Smith and each manifested itself in a way unique to Detroit.

The approach here is less about cause and effect, however, and more about sound/noise as a means of investigating complex interrelated/intercausal phenomena. Smith's argument demonstrates an approach similar bur perhaps less formal than the sonic economy being argued for here. She says:

> To fully understand Motown's relationship to the civil rights movement in all its complexity, one must begin on the streets of Detroit. Here, amateur singers, autoworkers, preachers, musicians, poets, business leaders, community activists, and politicians all participated in an elaborate 'dance' to reclaim the city for African Americans. The dancers moved in unpredictable patterns but shared many of the same steps. For some, participation involved seeking public office or organizing community action groups. For others, it involved publishing a newspaper, broadcasting a radio show, or singing a song. For still others, it involved opening a business, forming a union, or buying a home. In total Detroit's black community choreographed a distinctly urban social movement.
>
> (Smith 1999: 19)

Choreographed or not – the unpredictable patterns that the dancers formed were significantly noisy to have an impact, and despite the assertion from Harold Cruse in his *The Crisis of the Black Intellectual* (1967), that such a significant event, that by necessity combined political, economic and cultural force could only occur in Harlem, it happened in Detroit. This is important in terms of predictability and noise and chaos: no matter how much is known about the potential variables of any given situation, something other than that which was predicted is likely to actually occur.

As the 1970s approached their end, there were winds of change in the air. All that was solid was seemingly about to melt into air. Old certainties were being questioned and debates relating to turns, divergences, discontinuities, and new beginnings, new times – postmodern/post-industrial, etc. were rife. These changing times could be heard in the sound of Detroit, and specifically in the sounds of Juan Atkins, Derrick May, and Kevin Saunderson. The sound of the city shifted from the popular soul of Motown and the white rock music that moved from the MC5, through the Stooges to Alice Cooper and Kiss, to a new machine sound that seemed to usher in the computer age as the wheels of industry slowed, as in Coventry, almost to a halt. To these new producers synthesizers defined their city. 'They rang futuristic and aggressive, monotonous and melancholic. All this comprised the essence of Detroit techno' (Kilian Hanf 2010: 30).

The sound was a mixture, as Derek May acknowledged, of P Funk and Kraftwerk, it was a peculiar assemblage that drew on the past, both locally and globally, while simultaneously pointing to the future. It took the aesthetic of a city in decline, with Derek May describing Detroit, as Jerry Dammers had done before him in relation to Coventry, as a 'ghost town', and set it to a rhythm derived from the traditions of German Expressionism in films like *Metropolis*, and the musical experimentalism of Stockhausen, and later Düsseldorf's Kraftwerk.[10] It fused these echoes with a vision of the future that also drew on the ideas of futurologist Alvin Toffler, and Derek May was again explicit in his formulation of techno as part of the Third Wave of post-industrial development.[11]

For May, techno was the sound of industrial decline, and of a radical structural turn in economic terms, but didn't we say the sound of The Specials was the sound of decline? How can they both be, when they sound so very different? Because the sound is not representational, but a creative element within a complex network of other related factors of political economy that together produce a unique event. The point is not what the sound was but that a sound was made. The sound was not caused by economic decline as a by-product, the sound was in the event, and in both cases, it was linked to creative individuals also operating in relation to the event that was different as repeating and entangled singularity, and it was a repetition that operated as a transgression putting laws into question in an environment where authority and dissent occupied the same space (Deleuze 2008: 3).

The sound of Detroit techno soon found a global audience, and it found a particularly welcome home in the rave scene that was developing in the UK around the mid-1980s. The rock reggae fusion of The Specials described above was dispersing and becoming less audible as a different kind of noise began to emerge. Ironically, in Detroit, the export of techno to the rest of the world meant that it became less audible there as its key exponents dispersed to reap the rewards of their growing international popularity. But Detroit continued to produce its own unique noise, and in the absence of May, Atkins and Saunderson, a new noise-event began to emerge from the underground.[12] This new brand of techno, like its predecessor, and like the emergence of The Specials, demonstrated a particular kind of political, economic and aesthetic energy.

This second wave of techno, which continues up to today, was spearheaded by the anonymous and fiercely radical Underground Resistance (UR) who, in the absence of their forebears, set out to defend the integrity of Detroit from the vagaries of the commercial music industry. Underground Resistance was established by Jeff Mills and Mike Banks as a multifaceted and secret platform from which to launch sonically infused political outbursts and locally engaged political programmes.[13] As such, it constituted an expressive force, in the same way as the work of The Specials had done in Coventry. It was the same difference that repeats but does not resemble.[14]

The ambiguity in terms of the identity of UR's members, like the name The Special AKA, a name that always opens up the possibility of an alternative identity, is significant in terms of non-representation and the notion of the unsaid, the hidden, as set out in Chapter 3 by Maitland. As such, it constitutes an energy that does not conform to the ordering mechanisms of representational thought, and is therefore not constrained by the common-sense laws that form the basis of a traditional political economy. Static representational thinking may cite certain characteristics as defining features of an organized and predictable taxonomy, for example, physical geography of the kind reported by Marco Polo in Chapter 3, or economic history: I could state how many factories produced how many cars in a given time period in Coventry or Detroit, tell you how many people worked there and how much they got paid. I could tell you how many were black and how many were white, how many were male how many female, etc. I could tell you, as I have already done, how far it is between Coventry and Detroit or that it is 4,620 miles from Kingston to Coventry, or 4,021miles Düsseldorf to Detroit.

To an extent this would be useful and maybe even interesting information. And these are not insignificant factors: representational elements comprise essential elements even in a non-representational nexus, by which I mean Coventry and Detroit are aware of each other – as the opening examples demonstrated – and their shared characteristics or resemblances. These conscious elements are always present but not always visible. They form part of the reason for the comparative study being elaborated here. But they also stand to be critiqued and to point as much to difference as they do to repetition. In a chaotic environment, the unfolding uncertainty must always figure, as must the possibility for linkages beyond the simply quantifiable.

What is at issue here, then, is something quite different from the static ontic logic of representational resemblance. What are being described are coherences, and intensities in space, but intensities that insofar as they are always becoming are also events.

In that case Coventry and Detroit are not uniquely similar, without the intervention of conscious human agency, but given that conscious human agency is regarded here as a factor in a non-hierarchical quantum plenum, then there is a bond, as event, an entanglement. This point contrasts with the argument formulated by Meillassoux in Chapter 3, and by speculative realism or object-orientated philosophy more generally, and suggests more of an alliance with Stephen Shaviro in his *Without Criteria*, in which he develops the argument of A.N. Whitehead as follows:

> In *Process and Reality*, Whitehead strictly distinguishes between occasions and events, and between entities and societies. He 'uses the term "event" in the more general sense of a nexus of actual occasions, inter-related in some determinate fashion in one extensive quantum. An actual occasion is the limiting type of an event with only one member' (Whitehead: 73). At the limit, an event may be just one particular occasion, a single incident of becoming. But more generally, it is a group of such incidents, a multiplicity of becomings: what Whitehead calls a nexus. A nexus is 'a particular fact of togetherness among actual entities' (Whitehead: 20); that is to say, it is a mathematical set of occasions, contiguous in space and time, or otherwise adhering to one another. When the elements of a nexus are united, not just by contiguity, but also by a 'defining characteristic' that is common to all of them, and that they have all 'inherited' from one another, or acquired by a common process, then Whitehead calls it a society (Whitehead: 34). A society is 'self-sustaining; in other words . . . it is its own reason . . . The actual

> things that endure,' and that we encounter in everyday experience, 'all are societies' (Whitehead 1967: 203–204).
>
> (Shaviro 2012: 18)

This rather long quote needs to be unpicked so that its relevance here is clear.

An event is the connection of occasions that when joined form one extensive quantum: a singularity, an energy that is formed from 'a particular fact of togetherness'. This togetherness, or 'mathematical set' can be identified as contiguous or as having a 'defining characteristic' in common that is 'inherited' each from the other. The nexus being described here in relation to Coventry and Detroit is of the latter kind – not always contiguous in space and time, but a musical/sonic inheritance of common characteristics (as discussed in Chapter 3 in relation to the work of Leroi and Swire). It is 'a particular fact of togetherness among actual entities' (Shaviro 2012: 20).

If we are going to suggest that Coventry and Detroit share something in common, then it must be at the level of occasion/event/society that can take up quantum superpositions.

In both Coventry and Detroit, mobile populations operated as contributing elements in a process that fused music that had been dispersed with those populations with localized phenomena to produce spatially specific noise. In both cases: 'The contagious vibrations, sonic processes, and market tactics of strains of popular music within the African diaspora both extend the concept of an audio virology and offer a tactical outline of an affective *mobilization* as opposed to the *modulation* of preemptive capital' (Goodman 2010: 156).

What occurred in both cities was not a modulation between boom and bust, between industrial and post-industrial, but a mobile process of continuous change punctuated by discontinuous events. What repeated in both cases was an energy (like a festival) and individual humans were a part of that – Dammers, May, *et al.*, but not in terms of their individual identity, but as an event – as factors in the creation of noise. This repeating singularity can pop up in various states and in various places, sometimes simultaneously, as a strange entanglement. So Coventry and Detroit are entangled, rather than similar.[15]

The story of post-industrial economic restructuring, so well embodied in the fortunes of Coventry and Detroit and in the term 'Motorcity' with its conflicting emblems of movement and stasis or decay, and the subsequent drift into a new or significantly different, or digital paradigm, has not yet been adequately accounted for. The all too brief attempt to encapsulate some of this story above showed that comparing two places according to formal representational or visual measures will bring certain, quite predictable but largely uninteresting results. The resemblances between Coventry and Detroit that can be visually verified are at best limited. Demonstrating this points to the need for a much deeper and more detailed sonic investigation than has been possible here. But the results of such an enquiry can be speculated upon in relation to the arguments that have been developed so far in this book.

It is a story best told without beginning or end and through sonic means where visualization fades to reveal intrigue, betrayal, political action, heroes and villains, all set to a chaotic soundtrack.

As Brandon LaBelle has said: 'The seemingly innocent trajectory of sound as it moves from its source and toward a listener, without forgetting all the surfaces, bodies, and other sounds it brushes against, is a story imparting a great deal of information fully charged with geographic, social, psychological, and emotional energy' (Labelle 2010: 2). Both Coventry and Detroit are emblematic of industrial decline and of rebirth, and of the dynamic flow of people in and out. And they share certain characteristics. But examining the two places together exposes the limitations of representational thinking. By imposing representational similarities, we begin to get a number of different markers that can be used to measure and test certain cause-and-effect relationships. Once we start to look beyond such markers and turn to less quantifiable data, such as music, for example (that both Coventry and Detroit have a specific musical sound is one thing, but to listen to it is another), the situations become more difficult to make sense of and simple comparisons seem inadequate in terms of their ability to account for complex patterns of events. The need for this reinvigorated kind of thinking has been made greater by the decentred multiplicity of the digital age where the flow in and out of cultural exchange and hybridity have risen to unprecedented levels.

Both examples demonstrate an opening out rather than closing in. Pushing the closed signifiers of cars, race and music to their limits and beyond shows how rigid representational thinking can only go so far in accounting for the complexity of issues present in any contemporary socio-economic system. This comparison takes immediate and obvious similarities and, rather than treating them as closed signifiers, opens them out, extends them into non-representational realms where time and space are stretched. It addresses the relationship between global and local phenomena (or stresses), referencing Coventry and Detroit where the populations have experienced economic restructuring following the decline of their respective car industries.

We are using the concept of 'sonic economy' not only as a means to extend and challenge existing methodologies but also to re-centre an investigation of the processes at work in the locale, producing reactions to economic restructuring. The aim has been to identify locally induced innovations and changes which, on the one hand, are reactions to economic restructuring but, on the other (and importantly), draw on the redirected energy and latent potential of the affected population to produce new forms of (personal) investment, a new direction of labour, new constructions of value (which are socially and culturally embedded) all of them expressed in forms and places of production and consumption.

Faced with way-of-life challenging change, those who were not involved in 'flight' engaged or re-engaged with their environment in new ways in order to sustain themselves. At Motown, Two-Tone, and UR, a particular kind of political economy played itself out – a kind that resonates between and conjoins the two cities.

In an age of increasing technological complexity, the ability to locate and focus on important sites of political activity is becoming problematic. It is no longer clear where to focus because dispersion is so much a feature of the mediated environment. Hence it is more important than ever to rethink what we mean when we talk of the political and to restage it in a non-representational manner that embraces play and festival, and the sounds of Coventry and Detroit as significant sites of political enquiry. To be able to do this has required a new methodology – one capable of accounting for the dispersed and consequently invisible phenomena that cannot simply be named and grouped because they are constantly shifting in relation to one another. This need not imply chaos of the kind described by Adorno in the Introduction to this book, but chaos in the scientific sense of fractals and feedback loops where patterns emerge and can be identified according to a sonic rather than a visual logic.

The chapter, and this book for that matter, have been comprised of three synchronous aspects, each continually informing and responding to the others.

The first aspect traced the trajectories of two specific musical movements as they played out among the political and economic debris of Coventry and Detroit. It produced a genealogy (albeit limited) of each case within its own political, temporal, spatial and economic context. It concentrated on the Two-Tone phenomenon in Coventry where Jamaican ska resonated with the prevailing social and political climate, and Motown and Techno in Detroit, where music that flowed from diverse sources was adopted and adapted in specific and localized ways.

The second aspect examined the concepts of identity and difference, posing questions as to the efficacy of representational thinking and its ability to explain contemporary political phenomena. It addressed the idea that Coventry and Detroit resonate somehow. Here specific similarities and differences have been exposed to critical analysis: the point has been to identify patterns that emerge across space and time that might cite Coventry and Detroit as similar in terms of motor production, industrial and urban decline, and issues of racial conflict, and yet different and unique in terms of sound. This is important because the sounds that resonate in certain specific political locales or milieux can tell us something significant about what is happening in those places and beyond. Carrying out this comparison has shown how we can find significant details that are revealed through listening that would not have been deduced from visual representative, statistical, enquiry.

The third aspect has been to situate these specific examples in relation to a philosophical trajectory intended to aid us in an assessment of the importance of music and sound more generally within the contemporary political, social, and economic contexts. Moving beyond the dialectical markers of opposition and negation, to embrace Deleuze and Guattari's notions of the refrain and affirmation as fluid mechanisms for managing the seemingly unmanageable, a sense of how one can use sound as an organizing principle is methodologically established. The intention has been to develop a spatial and temporal sense of noise/sound/music as a political force that draws on authors and thinkers such as Heidegger, Bergson, Bachelard, Foucault, Lefebvre, Attali, Thrift, and Oliveros, to arrange political, economic, technological and cultural factors, into harmonious scores with diverse statements organized into discourses or agendas.

Chaos Media has been witness to troubled times in two motor cities emblematic of modern industrial production and the technological turn over the past 30-year period. The telling of their story uncovers pertinent factors not just about history, movement, and how we got to this point, but allows for a more comprehensive account of the contemporary moment and how it relates to the past as well as the future. Events resonate simultaneously and also sequentially. Arrangements – political and economic – can also be assessed in this way, drawing patterns in space and time to explain without recourse to crude cause-and-effect models that rely so heavily on repetition and predictability. As the urban spaces in question have become reconfigured to form post-industrial landscapes with intricate networks of connection and exchange, they have done so in a manner that defies explanation in traditional terms. The formal pillars of political economy alone cannot adequately account for the complexities of this post-industrial digital age of seemingly irregular patterns of coded exchange.

How then can restaging political economy within a sonic environment allow for a clearer understanding of contemporary mediated spaces that appear to be in a constant state of flux/drift? The answer lies in the proposition of a mode of thinking that simultaneously brings areas of life often regarded as peripheral (noise/sound/music) to the fore, and exposes traditionally important spheres of life (politics and economics) to sonic and not just visual enquiry. In doing so, it challenges approaches based on a straightforward linear narrative. Visual logic too often points to dialectical movement, with the digital replacing analogue, post-industrial replacing industrial, and computer music replacing soul. But a sonic economy refutes this and instead proposes a virtual plane on which statements as events emerge from and recede back into the ground – in and out of noise – as discontinuous continuity. The digital age did not enact a radical break with its analogue predecessor. The phenomenon may be cited as a discontinuous moment, but it was one that occurred within a continuous and enduring state of chaos from which patterns can and do emerge.

Notes

1 See http://ripitupfootnotes.blogspot.com/2008/11/footnotes-16-chapter-15-ghost-dance-2.html [Accessed 10 Sept. 2014].

2 See http://www.modeldmedia.com/features/detroitcoventry19709.aspx [Accessed 10 Sept. 2014].

3 See http://www.coventrytelegraph.net/lifestyle/nostalgia/look-coventry-transformed-britains-motor-6742407 [Accessed 10 Sept. 2014].

4 See http://www.coventry.gov.uk/info/448/records_and_archives-information_and_advice/356/local_history_and_archives/5 [Accessed 10 Sept. 2014].

5 See Healey and Clark (1984).

6 See http://www.fantazia.org.uk/Scene/orgs/eclipse.htm [Accessed 10 Sept. 2014].

7 Like Bachelard's creative moment, these voices were formed of echoes that had been reverberating in space until they collided with other elements that amplified them anew. This newness is significant in terms of the debate around continuity and creativity as expounded by Bergson and Bachelard respectively.

8 The decline of Detroit is explained in: http://www.bbc.co.uk/news/world-us-canada-23369573 [Accessed 10 Sept. 2014].

9 See http://www.autolife.umd.umich.edu/Race/R_Overview/R_Overview1.htm [Accessed 10 Sept. 2014].

10 For a more detailed account of the significance of Krartwerk in relation to Detroit Techno, see Kilian Hanf (2010)

11 Derek May said of Detroit:

> This City is in total devastation. It is going through the biggest change in its history. Detroit is passing through its third wave, a social dynamic which nobody outside this city can understand. Factories are closing, people are drifting away, and kids are killing each other for fun. The whole order has broken down. If our music is a soundtrack to all that, I hope it makes people understand with what kind of disintegration we're dealing with.
>
> (Cosgrove 1988: 89)

12 For a detailed account of the rise of techno and its movement out of Detroit onto the international stage, see Dan Sicko's *Techno Rebels: The Renegades of Electronic Funk* (Detroit: Wayne State University Press, 2010).

13 Steve Goodman writes that for Underground Resistance 'The city of Detroit becomes a vast rhythm machine, with mechanically pulsed affective waves rippling

intensity across the urban skin, carrying sonic parasites to hijack your nervous system' (2012: 157).

14 For a more detailed account of Underground Resistance, see Dan Sicko, *Techno Rebels* (2010).

15 On New Year's Eve 1981 – Sun Ra performed Auld Lang Syne in Detroit – Jerry Dammers would later reprise the Arkestra in his own style. See http://theworldsamess.blogspot.co.uk/2013/12/auld-lang-syne-sun-ra-in-detroit-198182.html [Accessed 10 Sept. 2014].

Bibliography

Adil, A. and Kennedy, S. (2009) 'Technology on Screen: Projection, Paranoia and Discursive Practice', in F J. Ricardo (ed.), *CyberCulture and New Media*, Amsterdam: Rodopi, pp. 219–30.

Adorno, T. and Horkheimer, M. (1997) *Dialectic of Enlightenment*, London: Verso.

Angell, I. (1995) 'Winners and Losers in the Information Age', *LSE Magazine*, Summer, 7(1).

Aranowitz, S., Martinsons, B. and Menser, M. (1996) *Techno Science and Cyberculture*, London: Routledge.

Arthur, B. (1999) Available at: http://tuvalu.santafe.edu/~wbarthur/Papers/EconandComplex_Web.pdf [Accessed 10 Sept. 2014].

Atkinson, Q.D. and Gray, R.D. (2005) 'Curious Parallels and Curious Connections: Phylogenetic Thinking in Biology and Historical Linguistics', *Systematic Biology*, 54: 513–26.

Attali, J. (2006) *Noise: The Political Economy of Music*, Minneapolis, MN: University of Minnesota Press.

Axford, B. and Huggins, R, (2006) *New Media and Politics*, London: Sage.

Bachelard, G. (1994) *The Poetics of Space*, Boston: Beacon Press.

Back, L. (2007) *Listening*, Oxford: Berg.

Baudrillard, J. (1983) *Simulations*, New York: Semiotext(e).

Bell, D. (1974) *The Coming of Postindustrial Society*, London: Heinemann.

Benjamin, W. (1936) Available at: http://www.marxists.org/reference/subject/philosophy/works/ge/benjamin.htm [Accessed 10 Sept. 2014].

Bergson, H. (1992) *The Creative Mind: An Introduction to Metaphysics*, New York: Citadel.

Biddle, I. and Knights, V. (2013) *Music, National Identity and the Politics of Location: Between the Global and the Local*, Burlington, VA: Ashgate.

Blesser, B. and Salter, L.R. (2007) *Spaces Speak, Are You Listening? Experiencing Aural Architecture*, Cambridge, MA: MIT Press.

Bolter, W.G., McConnaughey, J.W. and Kelsey, F.J. (1990) *Telecommunications Policy for the 1990s and Beyond*, New York: M.E. Sharpe.

Boyd, R. and Richerson, P.J. (1985) *Culture and the Evolutionary Process*, Chicago: University of Chicago Press.

Brett, P. (ed.) (1994) *Queering the Pitch: The New Gay and Lesbian Musicology*, New York: Routledge.

Bull, M. (2007) *Sound Moves: iPod Culture and Urban Experience*, London: Routledge,

Bull, M. and Back, L. (2013) *The Auditory Culture Reader (Sensory Formations)*, Oxford: Berg Publishers.

Callinicos, A. (1991) *Against Postmodernism: A Marxist Critique*, Cambridge: Polity Press.

Calvino, I. (1978) *Invisible Cities*, New York: Harcourt Brace Jovanovich.

Camus, A. (1965) *The Rebel*, London: Penguin.

Carey, J. and Elton, M.C.J. (2010) *When Media Are New: Understanding the Dynamics of New Media Adoption and Use*, Ann Arbor, MI: University of Michigan Press.

Casey, E.S. (1998) *The Fate of Place: A Philosophical History*, Los Angeles: University of California Press.

Castells, M. (1989) *The Informational City: Information Technology, Economic Restructuring, and the Urban Regional Process*, Oxford: Blackwell.

Cavalli-Sforza, L.L. and Feldman, M.W. (1981) *Cultural Transmission and Evolution: A Quantitative Approach*, Princeton, NJ: Princeton University Press.

Clark, D. (1989) *Urban Economic Decline*, London: Routledge.

Collins, R. and Murroni, C. (1996) *New Media, New Policies*, Cambridge: Polity Press.

Command Paper (1994) *Creating the Superhighways of the Future: Developing Broadband Communications*, CM 2734, London: HMSO.

Connell, J. and Gibson, C. (2003) *Sound Tracks: Popular Music, Identity, and Place*, London: Routledge.

Cosgrove, S. (1988) 'Seventh City Techno', *The Face*, 97 (May 1988): 86.

Cruse, H. (1967) *The Crisis of the Negro Intellectual: A Historical Analysis of the Failure of Black Leadership*, New York: New York Review Books.

Daniels, M.D. (2002) 'Theory Versus Analytics of Power: The Habermas–Foucault Debate'. Available at: http://www.csudh.edu/dearhabermas/habfouc02.htm [Accessed 10 Sept. 2014].

Darwin, C.R. ([1882] 1981) *The Descent of Man and Selection in Relation to Sex*, Princeton, NJ: Princeton University Press.

Deleuze, G. (1995) *Negotiations*, New York: Columbia University Press.

Deleuze, G. (1999) *Foucault*, Minneapolis, MN: University of Minnesota Press.

Deleuze, G. (2006) *The Fold: Leibniz and the Baroque*, London: Continuum.

Deleuze, G. (2008) *Difference and Repetition*, London: Continuum.

Deleuze, G. and Guattari, F. (1987) *A Thousand Plateaus: Capitalism and Schizophrenia*, trans. B. Massumi, Minneapolis, MN: University Minnesota Press.

Department of Trade and Industry (1990) *Competition and Choice: Telecommunications Policy for the 1990's*, White Paper, Cm1461, London: HMSO.

Derrida, J. (1978) *Writing and Difference*, London: Routledge.

Diffey, T.J. (2004) 'On Steering Clear of Creativity', *Journal of Visual Art Practice*, 3(2): 91–102.

Dunn, M., *et al.* (2005) 'Structural Phylogenetics and the Reconstruction of Ancient Language History', *Science*, 309(2): 1072–5.

Dyson, F. (2009) *Sounding New Media: Immersion and Embodiment in the Arts and Culture*, Berkeley, CA: University of California Press.

Finnegan, F., Salaman, G. and Thompson, K. (1987) *Information Technology: Social Issues*, London: Open University Press.

Fischer, H. (1997) 'A Framework for Global Electronic Commerce', *White House*, 1 July 1997. Available at: http://www.pff.org/issuespubs/futureinsights/fi6.1globaleconomiccommerce.html [Accessed 10 Sept. 2014].

Fischer, H. (2006) *Digital Shock: Confronting the New Reality*, Montreal: McGill-Queen's University Press.

Floridi, L. (2009) *Synthese*, 168: 151–78. Available at: http://philsci-archive.pitt.edu/4076/1/ado.pdf [Accessed 10 Sept. 2014].

Foucault, M. (2004a) *The Archaeology of Knowledge*, London: Routledge.

Foucault, M. (2004b) *The Order of Things*, London: Routledge.

Foucault, M. (2004c) *Death and the Labyrinth*,. London: Continuum.

Fukuyama, F. (1989) Available at: http://www.kropfpolisci.com/exceptionalism.fukuyama.pdf [Accessed 10 Sept. 2014].

Fuller, M. (2007) *Media Ecologies: Materialist Energies in Art and Technoculture*, Cambridge, MA: MIT Press.

Fuller, M. and Goffey, A. (2012) *Evil Media*, Cambridge, MA: MIT Press.

Gaffne, P. (2010) *The Force of the Virtual: Deleuze, Science, and Philosophy*, Minneapolis, MN: University of Minnesota Press.

Gates, B. (1996) *The Road Ahead*, London: Penguin.

Goddard, M., Halligan, B. and Hegarty, P. (2012) *Reverberations: The Philosophy, Aesthetics and Politics of Noise*, London: Bloomsbury.

Golding, M. (2011) 'Fractal Philosophy', *C Theory*. Available at: http://www.ctheory.net/articles.aspx?id=634 [Accessed 10 Sept. 2014].

Goodman, S. (2010) *Sonic Warfare: Sound, Affect, and the Ecology of Fear*, Cambridge, MA: MIT Press.

Grauer, V.A. (2006) 'Echoes of Our Forgotten Ancestors', *The World of Music*, 48(2): 5–59.

Gray, R.D. and Atkinson, Q.D. (2003) 'Language-Tree Divergence Times Support the Anatolian Theory of Indo-European Origin', *Nature*, 426: 435–9.

Gray, R.D. and Jordan, F.M. (2000) 'Language Trees Support the Express-Train Sequence of Austronesian Expansion', *Nature*, 405: 1052–5.

Habermas, J. (1979) *Communication, Evolution and Society*, London: Heinemann.

Habermas, J. (1992) *The Theory of Communicative Action*, Cambridge: Polity Press.

Hainge, G. (2013) *Noise Matters: Towards an Ontology of Noise*, London: Bloomsbury.

Haraway, D. (1991) 'Situated Knowledges', in *Simians, Cyborgs, and Women*, New York: Routledge.

Harman, G. (2009) *Prince of Networks: Bruno Latour and Metaphysics*, Melbourne: re.press.

Harman, G. (2010) *Towards Speculative Realism: Essays and Lectures*, Winchester: Zero Books.

Harman, G. (2011) *The Quadruple Object*, Winchester: Zero Books.

Harvey, D. (2012) *Rebel Cities: From the Right to the City to the Urban Revolution*, London: Verso Books.

Hayles, K. (1999) *How We Became Posthuman: Virtual Bodies in Cybernetics, Literature, and Informatics*, Chicago: University of Chicago Press.

Healey, M. and Clark, D. (1984) 'Industrial Decline and Government Response in the West Midlands: The Case of Coventry', *Regional Studies*, 18(4): 303–18.

Heap, N., Thomas, R., Einon, G., Mason, R. and Mackay, H. (eds) (1995) *Information Technology and Society*, London: Sage.

Heidegger, M. (1977) *The Question Concerning Technology and Other Essays*, New York: Harper and Row.

Heller, A. and Feher, F. (1988) *The Postmodern Political Condition*, New York: Columbia University Press.

Hillis, K. Available at: http://www.unc.edu/~khillis/techrep.html [Accessed 10 Sept. 2014].

Hillis, K. (1990) *Digital Sensations: Space, Identity, and Embodiment in Virtual Reality (Electronic Mediations)*, Minneapolis, MN: University of Minnesota Press.

Hollins, T. (1984) *Beyond Broadcasting: Into the Cable Age*, London: BFI.

Hunt Report (1982) *Inquiry into Cable Expansion and Broadcasting Policy*, Home Office Inquiry, London: The Stationery Office.

Ihde, D. (1993) *The Philosophy of Technology: An Introduction*, New York: Paragon Issues in Philosophy.

Ihde, D. (2001) *Bodies in Technology (Electronic Mediations)*, Minneapolis, MN: University of Minnesota Press.

Ihde, D. (2007) *Listening and Voice: Phenomenologies of Sound*, Albany, NY: State University of New York Press.

Information Technology Advisory Panel (1983) *Making a Business of Information: A Survey of New Opportunities*, London: The Cabinet Office.

Innis, H. (1951) *The Bias of Communication*, Toronto: University of Toronto Press.

Janus, A. (2011) 'Listening: Jean-Luc Nancy and the "Anti-Ocular" Turn in Continental Philosophy and Critical Theory', *Comparative Literature*, 63(2): 182–202.

Keiko, H. (2009) 'Bachelard's Theory of Time: Missing Link between Science and Art', *Aesthetics*, 13: 1–9.

Keizer, G. (2010) *The Unwanted Sound of Everything We Want: A Book about Noise*, New York: Public Affairs.

Kennedy, S. (2011) 'Sonic Economy', *C Theory*. Available at: http://www.ctheory.net/articles.aspx?id=649 [Accessed 10 Sept. 2014].

Khalfa, J. (ed.) (1999) *An Introduction to the Philosophy of Gilles Deleuze*, London: Continuum.

Kilian Hanf, M. (2010) *Detroit Techno: Transfer of the Soul through the Machine*, Berlin: Springer Verlag.

Kim-Cohen, S. (2009) *In the Blink of an Ear: Toward a Non-Cochlear Sonic Art*, London: Bloomsbury.

Kirby, A.D. (2003) 'Science Advisors, Representation and Hollywood Films', *Molecular Interventions*, 3:54–60.

Kirtzman, L.D. (ed.) (1988) *Foucault: Politics, Philosophy, Culture*, London: Routledge.

Kittler, F. (1995) 'There Is No Software', *C Theory*. Available at: http://www.ctheory.net/articles.aspx?id=74 [Accessed 10 Sept. 2014].

Kittler, F. (1999) *Gramophone, Film, Typewriter*, Redwood CA: Stanford University Press.

Kosko, B. (2006) *Noise*, London: Viking Penguin.

Krims, A. (2007) *Music and Urban Geography*, London: Routledge.

LaBelle, B. (2010) *Acoustic Territories: Sound, Culture and Everyday Life*, London: Bloomsbury.

Laclau, E. and Mouffe, C. (2001) *Hegemony and Socialist Strategy: Towards a Radical Democratic Politics*, London: Verso.

Latour, B. (2005) *Reassembling the Social: An Introduction to Actor Network Theory*, Oxford: Clarendon Press.

Lefebvre, H. (2004) *Rhythmanalysis*, London: Continuum.

Leibniz, G. (1992) *Discourse on Metaphysics and Other Essays: Discourse on Metaphysics; On the Ultimate Origination of Things; Preface to the New Essays; The Monadology*, Cambridge, MA: Hackett Publishing.

Leroi, A.M. and Swire, J. (2006) 'The Recovery of the Past', *The World of Music*, 48: 43–54.

Leyshon, A., Matless, D. and Revill, G. (1998) *The Place of Music (Mappings: Society, Theory, Space)*, New York: Guilford Press.

Lipsitz, G. (1997) *Dangerous Crossroads: Popular Music, Postmodernism and the Poetics of Place*. London: Verso.

Lomax, A. *et al.* (1968) *Folk Song Style and Culture*. Washington, DC: American Association for the Advancement of Science.

Lunenfeld, P. (1999) *Digital Dialectic: New Essays on New Media*, Cambridge, MA: MIT Press.

Lynch, K. (1960) *The Image of the City*, Cambridge, MA: MIT Press.

Lyon, D. (1988) *The Information Society: Issues and Illusions*, London: Polity Press.

Lyotard, J.F. (1984) *The Postmodern Condition: A Report on Knowledge*, Minneapolis, MN: University of Minnesota Press.

Lyotard, J.F. (2001) *The Soundproof Room: Malraux's Anti-Aesthetics (Cultural Memory in the Present)*, Stanford, CA: Stanford University Press.

Maitland, S. (2009) *A Book of Silence: A Journey in Search of the Pleasures and Powers of Silence*, London: Granta.

Malaspina, C. (2012) 'The Noise Paradigm', in M. Goddard, B. Halligan, and P. Hegarty (eds), *Reverberations: The Philosophy, Aesthetics and Politics of Noise*, London: Continuum.

Massumi, B. (2002) 'The Superiority of the Analog', in *Parables for the Virtual*, Durham, NC: Duke University Press.

McEvoy, J.P. and Zarate, O. (2004) *Introducing Quantum Theory*, Cambridge: Icon Books.

McLary, S. (1994) 'Constructions of Subjectivity in Schubert's Music', in P. Brett (ed.), *Queering the Pitch: The New Gay and Lesbian Musicology*, New York: Routledge, pp. 205–35.

McLuhan, M. (1967) *The Medium Is the Message*, Harmondsworth: Penguin.

McLuhan, M. (1975) *Understanding Media: The Extensions of Man*, London: Routledge & Kegan Paul.

McLuhan, M. and McLuhan, E. (1988) *Laws of Media: The New Science*, Toronto: University of Toronto Press.

McLuhan, M. and McLuhan, E. (2007) *Laws of New Media: The New Science*, Toronto: University of Toronto Press.
Meillassoux, Q. (2012) *After Finitude: An Essay in the Necessity of Contingency*, London. Bloomsbury.
Murdock, G.P. (1967). *Ethnographic Atlas: A Summary*, Pittsburgh, PA: The University of Pittsburgh Press.
Nancy, J-L. (2007) *Listening*, trans. C. Mandell. New York: Fordham University Press.
Nechvatal, J. (2011) *Immersion into Noise*. London: Open Humanities Press.
Negreponte, N. (1995) *Being Digital*. London: Hodder and Stoughton.
Negrine, R. (1985) *Cable TV and the Future of Broadcasting*. London: Croom Helm.
Nietzsche, F. (1967) *The Birth of Tragedy*, trans. W. Kaufman, New York: Vintage.
Nietzsche, F. (1997) *Thus Spoke Zarathustra*, Ware, Hertfordshire: Wordsworth Editions Ltd.
Oliveros, P. (2005) *Deep Listening*, New York: iUniverse.
Pangle, T.L. (1993) *The Ennobling of Democracy: The Challenge of the Postmodern Age*, Baltimore, MD: Johns Hopkins University Press.
Patton, P. and Protevi, J. (eds) (2003) *Between Deleuze and Derrida*. London: Continuum.
Plato (1987) *The Republic*. London: Penguin Classic.
Poster, M. (1984) *Foucault, Marxism and History*, Cambridge: Polity Press.
Poster, M. (1990) *The Mode of Information*, Cambridge: Polity Press.
Poster, M. (1995) *The Second Media Age*, Cambridge: Polity Press.
Priest, E. (2013) *Boring Formless Nonsense: Experimental Music and the Aesthetics of Failure*, London: Bloomsbury.
Prochnik, G. (2011) *In Pursuit of Silence: Listening for Meaning in a World of Noise*, Garden City, NY: Doubleday.
Raboy, M. and Bruck, P.A. (1989) *Communication For and Against Democracy*, New York: Black Rose Books.
Rancière, J. (2002) 'The Aesthetic Revolution and Its Outcomes'. Available at: http://newleftreview.org/II/14/jacques-ranciere-the-aesthetic-revolution-and-its-outcomes [Accessed 10 Sept. 2014].
Roberts, L. (2012) *Mapping Cultures: Place, Practice, Performance*, Palgrave Macmillan ebook.
Roussel, R. (1977) *How I Wrote Certain of My Books*, New York: Sun.
Rutsky, R.L. (1999) *High Techne: Art and Technology from the Machine Aesthetic to the Posthuman*, Minneapolis, MN: University of Minnesota Press.
Schopenhauer, A. (1966) *The World as Will and Representation, in Two Volumes*, trans. E.F.J. Payne, New York: Dover Publication.

Shannon, C.E. and Weaver, W. (1949) *The Mathematical Theory of Communication*, Urbana, IL: University of Illinois Press.

Shaviro, S. (2012) *Without Criteria: Kant, Whitehead, Deleuze, and Aesthetics*, Cambridge MA: MIT Press.

Sicko, D. (2010) *Techno Rebels: The Renegades of Electronic Funk*, Detroit: Wayne State University Press.

Slack, J.D. (1984) *Communication Technologies and Society: Conceptions of Causality and the Politics of Technological Intervention*, New Jersey: Ablex.

Slack, J.D. (1987) *Ideology of the Information Age*, New Jersey: Ablex.

Smith, S.E. (1999) *Dancing in the Street: Motown and the Cultural Politics of Detroit*, Cambridge, MA: Harvard University Press.

Spreen, J.F. and Holloway, D. (2005) *Who Killed Detroit?: Other Cities Beware*, New York: iUniverse.

Sterne, J. (2003) *The Audible Past: Cultural Origins of Sound Reproduction*, Durham, NC: Duke University Press.

Sterne, J. (2012a) *The Sound Studies Reader*, London: Routledge.

Sterne, J. (2012b) *MP3: The Meaning of a Format (Sign, Storage, Transmission)*, Durham, NC: Duke University Press.

Stivale, C.J. (2005) *Gilles Deleuze: Key Concepts*, Chesham: Acumen.

Stock, J.P.J. (1996) *World Sound Matters: An Anthology of Music from Around the World*, London: Schott.

Stock, J.P.J. (2006) 'Clues from Our Present Peers? A Response to Victor Grauer', *The World of Music*, 48(2): 73–91.

Stokes, M. (1994) *Ethnicity, Identity and Music: The Musical Construction of Place*, Oxford: Berg.

Tanizaki, J. (2010) *In Praise of Shadows*, Frankfurt: Springer-Verlag.

Thompson, D. (2004) *Wheels Out of Gear: 2Tone, The Specials and a World in Flame*, London: Helter Skelter.

Thompson, M. and Biddle, I. (2013) *Sound, Music, Affect: Theorizing Sonic Experience*, London: Bloomsbury.

Thrift, N. (2008) *Non-Representational Theory: Space, Politics, Affect*, London: Routledge.

Thrift, N. and Crang, M. (2000) *Thinking Space*, London: Routledge.

Toop, D. (2011) *Sinister Resonance: The Mediumship of the Listener*, London: Bloomsbury.

Touraine, A. (1971) *The Postindustrial Society, Tomorrow's Social History: Classes, Conflicts, and Culture in the Programmed Society*, New York: Random House.

Trower, S. (2012) *Senses of Vibration: A History of the Pleasure and Pain of Sound*, London: Bloomsbury.

Vattimo, G. (1989) *The Transparent Society*, Cambridge: Polity.

Vendler, H. (1995) *The Breaking of Style*, Cambridge, MA: Harvard University Press.

Virilio, P. (2005) *The Information Bomb*, London: Verso.

Voegelin, S. (2010) *Listening to Noise and Silence: Towards a Philosophy of Sound Art*, London: Bloomsbury.

Waldrop, M.M. (1992) *Complexity: The Emerging Science at the Edge of Chaos*, London: Penguin.

Ward, K. (1989) *Mass Communications and the Modern World*, Basingstoke: Macmillan.

Whitehead, A.N. ([1933] 1967) *Adventures of Ideas*, New York: the Free Press.

Whitehead, A.N. (1978) *Process and Reality*, New York: the Free Press.

Whitely, S., Bennett, A. and Hawkins, S. (2005) *Music, Space and Place: Popular Music and Cultural Identity*, Burlington, VA: Ashgate.

Williams, Raymond (1990) *Television, Technology and Cultural Form*, London: Collins.

Williams, Roger (1971) *Politics and Technology*, London: Macmillan.

Winner, L. (1980) 'Do Artifacts Have Politics?' *Daedalus*, 109(1): 121–36.

Winner, L. (1993) 'Upon Opening the Black Box and Finding It Empty: Social Constructivism and the Philosophy of Technology', *Science, Technology, & Human Values*, 18: 362–78.

Wolff, R.P. (1988) *Moneybags Must Be So Lucky: On the Literary Structure of Capital*, Amherst. MA: University of Massachusetts Press.

Yourgrau, P. (2005) *A World Without Time: The Forgotten Legacy of Godel and Einstein*, London: Penguin.

Žižek, S. (2001) Welcome to the Desert of the Real', in *Re:constructions: Reflection on Humanity and Media after Tragedy*, 15 Sept. 2001. Available at: http://web.mit.edu/cms/reconstructions/interpretations/desertreal.html [Accessed 10 Sept. 2012].

Web addresses

http://anthem-group.net/2008/01/11/graham-harman-on-mcluhan-and-heidegger/ [Accessed 10 Sept. 2014].

http://www.anthonyhempell.com/papers/tetrad/concept.html [Accessed 10 Sept. 2014].

http://www.batteryoperated.net/mag_articles.htm [Accessed 10 Sept. 2014].

http://www.christianhubert.com/hypertext/plane_of_immanence.html [Accessed 10 Sept. 2014].

http://www.modeldmedia.com/features/detroitcoventry19709.aspx [Accessed 10 Sept. 2014].

http://www.rainerlinz.net/NMA/repr/Representation.html [Accessed 10 Sept. 2014].

http://users.rcn.com/bmetcalf.ma.ultranet/What%20is%20Univocity.htm [Accessed 10 Sept. 2014].

Index

www.ingramcontent.com/pod-product-compliance
Ingram Content Group UK Ltd.
Pitfield, Milton Keynes, MK11 3LW, UK
UKHW020124290926
15471UKWH00003B/277

9 781501 324420